Martin Leuthold

Characterization of Membrane Adsorbers for Contaminant Removal

Martin Leuthold

Characterization of Membrane Adsorbers for Contaminant Removal

Südwestdeutscher Verlag für Hochschulschriften

Impressum / Imprint

Bibliografische Information der Deutschen Nationalbibliothek: Die Deutsche Nationalbibliothek verzeichnet diese Publikation in der Deutschen Nationalbibliografie; detaillierte bibliografische Daten sind im Internet über http://dnb.d-nb.de abrufbar.
Alle in diesem Buch genannten Marken und Produktnamen unterliegen warenzeichen-, marken- oder patentrechtlichem Schutz bzw. sind Warenzeichen oder eingetragene Warenzeichen der jeweiligen Inhaber. Die Wiedergabe von Marken, Produktnamen, Gebrauchsnamen, Handelsnamen, Warenbezeichnungen u.s.w. in diesem Werk berechtigt auch ohne besondere Kennzeichnung nicht zu der Annahme, dass solche Namen im Sinne der Warenzeichen- und Markenschutzgesetzgebung als frei zu betrachten wären und daher von jedermann benutzt werden dürften.

Bibliographic information published by the Deutsche Nationalbibliothek: The Deutsche Nationalbibliothek lists this publication in the Deutsche Nationalbibliografie; detailed bibliographic data are available in the Internet at http://dnb.d-nb.de.
Any brand names and product names mentioned in this book are subject to trademark, brand or patent protection and are trademarks or registered trademarks of their respective holders. The use of brand names, product names, common names, trade names, product descriptions etc. even without a particular marking in this works is in no way to be construed to mean that such names may be regarded as unrestricted in respect of trademark and brand protection legislation and could thus be used by anyone.

Coverbild / Cover image: www.ingimage.com

Verlag / Publisher:
Südwestdeutscher Verlag für Hochschulschriften
ist ein Imprint der / is a trademark of
OmniScriptum GmbH & Co. KG
Heinrich-Böcking-Str. 6-8, 66121 Saarbrücken, Deutschland / Germany
Email: info@svh-verlag.de

Herstellung: siehe letzte Seite /
Printed at: see last page
ISBN: 978-3-8381-3720-9

Zugl. / Approved by: Hannover, Leibniz University, Diss., 2012

Copyright © 2014 OmniScriptum GmbH & Co. KG
Alle Rechte vorbehalten. / All rights reserved. Saarbrücken 2014

Contents

Abstract (English) 4

List of abbreviations 6

List of symbols 8

List of figures 11

List of tables 14

1 Introduction 15

2 Objectives 17

3 Theoretical backgrounds 19

   3.1 Design of biopharmaceutical downstream processes 19

   3.2 Contaminants in downstream processing 23

   3.3 Membrane adsorbers 27

   3.4 Bottlenecks in process development 31

4 Implementation of a 96-well HTPD platform for the characterization of membrane adsorbers 38

   4.1 Device and screening platform 38

      4.1.1 Development of a reusable 96-well device 38

      4.1.2 Automation 42

      4.1.3 Liquid-handling 45

4.2 Screening of membrane adsorbers *49*

    4.2.1 Heterogeneity of membrane adsorbers on a small-scale format 50

    4.2.2 Influence of the flow rate on the binding performance 52

    4.2.3 Applications in high-throughput format 59

    4.2.4 Scalability of binding performance to larger devices 63

4.3 Conclusion *70*

5 Investigation of contaminant removal using anion exchanger membranes at various process conditions **72**

    5.1 Model systems for studying chromatography in downstream processing 73

    5.2 Experimental design and evaluation of results *75*

    5.3 Binding performance of anion exchanger membrane adsorbers *79*

        5.3.1 Influences of salt and pH on the binding of model contaminants 80

        5.3.2 Effect of mono- and multivalent salt on the separation of molecules 91

    5.4 Comparison of the separation of molecules to larger device *97*

    5.5 Conclusion *98*

6 Summary and outlook **100**

7 References **103**

8 Appendix **114**

    8.1 Materials *114*

        8.1.1 Chemicals and biomolecules 114

        8.1.2 Buffers and buffer preparation 116

8.1.3 Equipment 120

8.1.4 Software 122

8.1.5 Preparation of Green Fluorescent Protein 123

8.1.6 Preparation of Bacteriophage ΦX174 124

8.2 Methods *126*

8.2.1 Photometric determination of protein and DNA 126

8.2.2 Phosphate detection 128

8.2.3 GFP assay 128

8.2.4 Phage assay 130

8.2.5 DNA assay 132

8.2.6 Endotoxin assay 135

8.2.7 Measurements using a FPLC system or a peristaltic pump 138

8.3 Supplementary results *140*

8.3.1 Determination of volume correction factor and pipetting error 140

8.3.2 Separation of DNA and GFP 141

Abstract

The production of active pharmaceutical ingredients is based increasingly upon biotechnological processes. Beside the target molecule there are a variety of undesirable substances that need to be removed. Often several chromatographic steps are used during purification. In addition to conventional chromatographic resins today different membrane-based adsorbers have been established. The composition and concentration of contaminants, as well as the operating conditions like pH and conductivity values, vary depending on the biological expression system, purification scheme and the chromatographic performance. Membrane adsorbers with high contaminant-removal capabilities for a large variety of impurities in a wide operational window of pH and conductivity are of great interest in biopharmaceutical production. The understanding of the interaction of contaminants and biomolecules with membrane adsorbers is a crucial aspect in the development of new membranes and purification steps in downstreaming. In order to find the most effective ligands and purification conditions as well as the understanding of the process extensive experimental process development is necessary.

To reduce the effort, an appropriate, application-oriented small-scale test method was established in this thesis and was evaluated for the investigation of membrane adsorbers. This enables the parallel characterization of properties of the adsorbent in order to reduce time and material consumption in membrane and process development. For this purpose a screening device for receiving the membrane material and an automated platform were developed, which can be used to represent typical chromatographic steps as well as for analytical and preparative purposes. This membrane adsorber screening platform was estimated regarding possible applications and the chromatographic performance was compared to larger devices. Following different anion exchangers were characterized in terms of performance for contaminant binding in a wide range of pH and conductivity. An anion exchanger with ammonium ligands was compared with a salt-tolerant type based on polyallylamine ligands using different models. New approaches were investigated which allow a separation of target molecules and contaminants at similar net charge of the protein surface, e.g. by the addition of small amounts of multivalent salts.

List of abbreviations

ANOVA	Analyse of variance
BIS-Tris	Bis(2-hydroxyethyl)-amino-tris(hydroxymethyl)-methane
BSA	Bovine serum albumin
CHES	N-Cyclohexyl-2-aminoethanesulfonic acid
CHO	Chinese hamster ovary cell
EDTA	Ethylenediaminetetraacetic acid
DNA	Desoxyribonucleic acid
DSP	Downstream processing
FDA	Food and Drug Administration
FPLC	Fast protein liquid chromatography
GFP	Green fluorescent protein
HCl	Hydrochloric acid
HCP	Host cell protein
HTPD	High-throughput process development
HTS	High-throughput screening
IEX	Ionic exchanger
LHP	Liquid handling parameter
LHS	Liquid handling system
MA	Membrane adsorber
MAb	Monoclonal antibody

MW	Molecular weight
NaAc	Sodium acetate
NA	Nucleic acids
NaOH	Sodium hydroxide
RNA	Ribonucleic acid
Tris	Tris(hydroxymethyl)aminomethane
UV	Ultraviolet

List of symbols

Symbol	Units	Description
A	[cm²]	Membrane area
$actual_i$	[µl]	Volume of pipetting
BC	[mg/cm²]	General binding capacity
BC_{FT}	[mg/cm²]	Binding capacity calculated by analysing the flow-through
BC_{El}	[mg/cm²]	Binding capacity calculated by analysing the elution
BT_{cont}	[%]	Breakthrough of the contaminant
BT_{tm}	[%]	Breakthrough of the target molecule
c_0	[mg/ml], [PFU/ml], [EU/ml]	General initial concentration
$c_{Flow\text{-}Through}$	[mg/ml]	Concentration in the flow-through fraction
c_i	[mg/ml], [PFU/ml], [EU/ml]	General concentration in a sample
c_{lim}	[PFU/ml]	Detection limit of phage titer
c_{Load}	[mg/ml]	Concentration in the initial solution
c_{Wash}	[mg/ml]	Concentration in the wash fraction
D	[-]	Dilution factor
d_m	[µm]	Thickness of membrane
E	[-]	Sum of weighting
e	[-]	Necessary number of measurements
$Flow_{MV/min}$	[MV/min]	Flow rate during filtration in relation to membrane volume
K	[-]	Equilibrium constant
k	[-]	Number of influencing variables
LRV	[-]	Log reduction value

Symbol	Units	Description
m	[mg], [kg]	General amount of protein
m_0	[mg]	Amount of protein loaded to the membrane
n	[-]	Number of measurements
n_v	[-]	Number of setpoint of influencing variables
$Onset_{Is}$	[s]	Measuring time for diluted initial solution (endotoxin assay)
$Onset_{Sample}$	[s]	Measuring time for a sample (endotoxin assay)
$Onset_{SampleSTD}$	[s]	Standardized measuring time for a sample (endotoxin assay)
$Onset_{STD}$	[s]	Measuring time for calibration curve (endotoxin assay)
P	[-]	Number of countable plaques
p	[mbar]	Pressure
q	[mg/cm²]	Binding capacity at equilibrium
q_{max}	[mg/cm²]	Maximum binding capacity
R	[%]	Recovery of target molecule
S	[-]	Selectivity
set_i	[µl]	Setpoint of pipetting volume
t	[min], [s], [h]	Time
t_{BT}	[min]	Time of the filtration
UV_{max}	[au], [mAU]	Extinction of light
$\dot{V}$	[ml/min]	Flow rate
$V_{Elution}$	[ml]	Volume of the elution fraction
V_{Load}	[ml]	Loaded volume
V_{Sample}	[ml]	Volume of the sample
V_V	[ml]	Void volume

Symbol	Units	Description
V_{Wash}	[ml]	Volume of the wash fraction
$Volcor_{new}$	[-]	Recalculated volume correction factor
$Volcor_{old}$	[-]	Volume correction factor
X_i	various	Measured value
μ	various	Arithmetic average
σ	various	Standard deviation
υ	[%]	Coefficient of variation
τ	[min]	Residence time

List of figures

Figure 1: Mode of operation in chromatography for contaminant removal 21

Figure 2: Typical downstream process schemes 22

Figure 3: Charge of contaminants and antibodies 31

Figure 4: Chromatography resin screening 34

Figure 5: Chromatographic steps 35

Figure 6: Use of screening in membrane chromatography 36

Figure 7: Homemade 96-well membrane holder construction 39

Figure 8: 3D CAD construction of the 96-well membrane holder 40

Figure 9: Prove of absence of lateral diffusion 41

Figure 10: Trial to detect cross-contamination using phosphate ions 42

Figure 11: The liquid-handling system 44

Figure 12: Circuit diagram vacuum station 44

Figure 13: Screening program architecture 45

Figure 14: Flow rate depending on differential pressure 53

Figure 15: Binding of protein depending on residence time 55

Figure 16: Binding of phages depending on residence time 56

Figure 17: Binding of endotoxin depending on residence time 57

Figure 18: Effect of filling the wells without backpressure 59

Figure 19: Example loading scheme for chromatofocusing and breakthrough curves 60

Figure 20: Generic breakthrough curve of a membrane adsorber 61

Figure 21: Schematic breakthrough curves at different buffer conditions 64

Figure 22: Comparison of protein binding 65

Figure 23: Comparison of relative values in protein binding 66

Figure 24: Comparison of phage binding 67

Figure 25: Comparison of the breakthrough behaviour 69

Figure 26: Example of a contour plot 77

Figure 27: Example of a sweet spot analysis 78

Figure 28: Influence of NaCl and pH on the binding of phages 81

Figure 29: Influence of NaCl and pH on the binding of endotoxin 82

Figure 30: Influence of NaCl and pH on DNA binding 83

Figure 31: Influence of NaCl and pH on BSA binding 84

Figure 32: Influence of phosphate and pH on DNA binding 85

Figure 33: Influence of phosphate, citrate and pH on binding BSA IEX1 87

Figure 34: Influence of phosphate, citrate and pH on binding BSA IEX2 89

Figure 35: Influence of phosphate and chloride on GFP binding 90

Figure 36: Separation of DNA and GFP 93

Figure 37: Sweet spot analysis separation DNA and GFP 95

Figure 38: Separation of phages and GFP 96

Figure 39: Separation of DNA and phages using phosphate 97

Figure 40: Route of process development using membrane adsorbers 100

Figure 41: Adjustment of pH for the different buffers 118

Figure 42: Conductivity depending on sodium chloride concentration 119

Figure 43: Determination of protein concentration using BCA 124

Figure 44: Calibration of BSA using the Tecan plate reader at 280 nm 127

Figure 45: Calibration of DNA using the Tecan plate reader at 260 nm 127

Figure 46: Fluorescence of GFP depending on the concentration 129

Figure 47: Florescence of GFP depending on pH and sodium chloride 130

Figure 48: Standard curve of DNA using the PicoGreen® assay 134

Figure 49: Influence of salt and pH on the PicoGreen® assay for calf thymus DNA 135

Figure 50: Influence of salt and pH on the PicoGreen® assay for salmon sperm DNA 135

Figure 51: Influence of salt and pH on the endotoxin assay 137

Figure 52: Evaluation of the linearity of the UV 280 nm signal by the ÄKTA prime 139

Figure 53: Separation of DNA and GFP at pH 6 141

Figure 54: Separation of DNA and GFP at pH 7 142

Figure 55: Separation of DNA and GFP at pH 9 143

List of tables

Table 1: Generic methods for contaminant detection and quantification 27

Table 2: Chromatographic ligands of membrane adsorbers and interactions 29

Table 3: High-throughput applications in resins and membrane based chromatography 37

Table 4: Liquid-handling Parameter (LHP) 47

Table 5: Evaluation of the variation of the measurement 51

Table 6: Chemicals 114

Table 7: Biomolecules 116

Table 8: Basic buffers 117

Table 9: Equipment 120

Table 10: Deviation of ΦX174 plaque assay 132

Table 11: Protocol for preparing PicoGreen® standard curve 133

Table 12: Volume correction factor and estimation of pipetting error 140

1 Introduction

The demand for biopharmaceuticals is steadily growing [1-3]. The main reasons are the ongoing demand and development of new therapeutic approaches and diagnostics using recombinant proteins like hormones, clotting factors, antibodies, growth factors or vaccines. For example, monoclonal antibodies are used for therapy of cancers, autoimmune diseases or diagnostics. Even in 2006 the number of approved antibodies was over 20 and more than 160 had been examined for launch [4]. The sales of monoclonal antibodies reached 30 billion US dollars in 2008 [5]. Furthermore, a high potential is attributed to recombinant viral vaccines for gene therapy or vaccination. For those applications probably a large amount of virus particles is required [6].

Despite of the trend of growing diversity and sales, the development of economic bioprocesses under shorter timelines is required [7, 8], because e.g. more stringent safety regulations increase the costs of clinical trials [1, 9]. Furthermore, a cost reduction could improve the access to emerging markets and drives the production of biopharmaceuticals for rare diseases.

For the production recombinant eukaryotic and prokaryotic expression systems are used. During the cultivation, including the growth of cells and product formation, additionally to the target molecule a multiplicity of contaminants occur. Host cell proteins, viruses, adventitious protein aggregates, endotoxins or nucleic acids could lead to undesired effects for the product itself or the patient. The nature, composition and quantity of components to be removed during downstream processing vary depending on the expression system, nutrient solution and process conditions used for cultivation. The therapeutic effectiveness or stability can decrease [10]. Furthermore, contaminants are responsible for adverse health effects after administration. This is caused by the contaminant itself or changed active agents. The immune reaction caused by the contaminating substances can lead to complications during the therapy [11, 12].

Introduction

Thus, for the approval of biopharmaceuticals, e.g., by the Food and Drug Administration or the European Agency for the Evaluation of Medicinal Products, all interfering impurities have to be reduced down to an acceptable level. Depending on the amount of dose, guidance values for the permitted amount of impurities per dose exist.

Following the cultivation and the cell harvest, downstream processing includes several steps to purify and concentrate the target molecule. Here chromatographic processes based on size, ionic exchange, affinity or hydrophobic interactions are essential procedures. The performance of these steps depends on the composition of contaminants as well as the window of process conditions like pH, conductivity or buffer composition. In addition to traditional chromatographic resins, in recent years further technologies using adsorptive materials were implemented in downstream processing. Membrane adsorbers distinguish themselves especially due to the high-throughput, the effective utilization of the binding capacities and performance in the binding of larger molecules compared to chromatographic resins [13-17].

Based on the aforementioned facts the present work focuses on the development and evaluation of fast test procedures to characterise membrane adsorbers for contaminant removal using small-scale approaches. This is the fundament for the improvement of knowledge in contaminant removal with membrane adsorbers, the development of new applications and process optimisation. The focus is on the investigation of anion exchangers for applications in downstream processing. The properties of a standard and new salt-tolerant membrane adsorber are compared.

2 Objectives

The improvement of the use of membrane adsorbers in pharmaceutical bioprocesses can be divided into two areas: the development of membranes and the improvement of process understanding including the knowledge about interaction between target molecule, contaminants and membrane adsorber. For the implementation and characterization of membranes the three major objects are of interest:

1) A suitable device was necessary to generate a better understanding of the performance of membranes. The primary requirements were adapted to the use in the early stage of membrane and process development. Typically only small amounts of new modified membrane material or process solutions containing the molecule of interest are available. To accelerate development a quick installation and a simultaneous measurement of several process conditions or composition of target molecules and contaminants should be possible. The amount of materials for the tests has to be as small as possible.

2) For the measurement of a broad operational window of conditions the complexity in providing appropriate process solutions is a drawback. Furthermore, the amount of work for sample analysis increases. Therefore, an automated platform was established to characterise membrane adsorbers. It enables the flexible preparation of typical process buffers and protein solutions. The platform carries out all the pipetting steps in order to mimic a real chromatographic process sequence. A 96-well plate reader for the analytics is directly adapted to the automated platform.

3) The screening system was used to define process conditions and membrane properties which are crucial for contaminant removal. The

focus was on the comparison of a well-established and a new salt-tolerant anion exchange membrane adsorber. The knowledge should improve the understanding of a useful operational window during a purification step identifying critical parameters and evaluate new polishing strategies to separate contaminants and target molecules.

3 Theoretical backgrounds

3.1 Design of biopharmaceutical downstream processes

In principle there is no general approach for a successful production process. Instead, several process steps have to be implemented due to the expression system used and necessary safety requirements. For the release and marketing of a given biopharmaceutical, regulatory authorities request the use of different technologies during downstream processing to ensure the reduction of critical impurities like viruses. The target molecule can be produced using different cell lines like bacteria, mammalian or insect cells [18, 19]. Furthermore, yeast or plant cells are used [20]. After cultivation cell harvest initiates the transition of the target molecule to downstream processing. The necessary steps needed for the cell harvest basically depends on the cellular location of the target. If the product is not secreted into the media by the cell or microorganism, a cell disruption prior the separation of cell and target is necessary. For the release of target molecule physical (grinding, decompression, freezing and thawing, ultrasound), chemical (treatment with acid or extraction with acetone) or biological methods (enzymes, phages) can be used [21]. To remove unwanted, preferably insoluble components centrifugation, flotation or microfiltration is applied. Also during cell harvest it is preferable to remove contaminants like HCP, endotoxins or DNA. Examples are the reduction by depth filter materials [22] or precipitation [23]. Therefore, such process steps get a smoother processing from cultivation to final downstream processing.

General aims of downstream processing are the separation of the target molecule (product) from contaminants, adjusting the required concentration and stabilisation of the drug substance and using a storage medium for final filling. It should be of the highest purity and recovery possible. For concentration and purification several steps are used which can have an impact on all aforementioned aims. In the following the most common techniques are described and evaluated in terms of operational performance. The assessment refers primarily to the use in a process scale.

(Ultra-) Centrifugation uses centrifugal forces to separate components by density [24]. This technique is rather expensive. Because of the limitations regarding scalability, large-scale production and continuous processing, process developers try to avoid this technology [25]. Like with precipitation, extraction or crystallisation, centrifugation requires a further step to separate the resulting phases. The pellet can be dissolved in a different buffer, which is required for further purification steps or stabilisation of the product.

Filtration is a major part of downstream processing. As a function of pore size microfiltration (> 0.1 µm) and ultrafiltration (0.01 – 0.1 µm) are common used techniques. Sterile filtration removes microorganisms while the product passes the filtration unit [26]. Cross-flow filtration separates dissolved components in a solution using membranes with a defined molecular cut-off. Depending on the size the target molecule is retained or passes the membrane. While during dead-end filtration a growing filter cake reduces flux, cross-flow filtration has the advantage that the filter cake is washed away by tangential flow over the filter area. In the case that the target molecule is retained, it is furthermore possible to increase its concentration in the retentate. A special embodiment of cross-flow filtration is diafiltration. The target molecule will be retained by the membrane while small molecules like contaminants, water and medium ingredients pass through the membrane. By adding new buffer to the circulating retentate it is possible to change the buffer [27]. To assure virus clearance, beside virus inactivation by ultraviolet radiation, antiviral agents or low pH, size exclusion based virus filtration can be used [28, 29].

Chromatography is a well-established step in the reduction of contaminants and/or the concentration of the product. In biopharmaceutical processes ion exchange, hydrophobic interaction and affinity are the most common separation mechanisms of chromatography. Here, two modes of operation are practical. In flow-through mode the target molecule passes through the chromatographic media (stationary phase) ideally without an interaction (binding). Whereas, contaminants interact (bind) to the chromatographic media and are retained. This mode of operation is often located near to the end of a downstream process. The second mode is bind and elute. The target molecule binds preferably to the stationary phase whereas most of the contaminants pass through. Non-bound components are removed with a washing step. Furthermore, bind and elute enables the removal of contaminants using specific washing steps or elution profiles. Finally, the use of a different buffer initiates the release of the bound target molecule by quenching or weakening binding forces. In contrast to flow-through mode, bind and elute additionally increases the product concentration by using a smaller amount of elution buffer compared to the volume of the applied solution of the target. While the reduction of contaminants, often also called purification or polishing step, can be done in flow-through or bind and elute mode, capture chromatography or capturing describes the binding of the target molecule to an adsorbent [13, 30, 31].

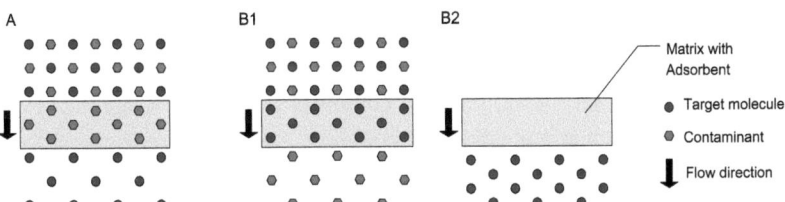

Figure 1: Mode of operation in chromatography for contaminant removal.
In flow-through (A) the filtrate ideally contains only the target molecule, while contaminants of the feedstream bind to the adsorbent. Bind and elute takes place in at least two steps: During loading (B1) the target molecule binds to the adsorbent, whereas the contaminants preferably pass through. In a second step (B2) the target molecule is removed by the displacement or weakening of the interaction.

Theoretical backgrounds

After purification and concentration, formulation by drying, crystallisation or final filling is used for preservation of the active drug substance [32].

The following flow charts illustrate two potential processes for the production of monoclonal antibodies (mAb) [17, 33-35] and viral vaccines [36]. Widespread mAbs are used in diagnostics and for therapeutic treatment [37], viral vaccines for immunization. Special forms of viral vaccines are virus-like particles which do not contain genetic material [38].

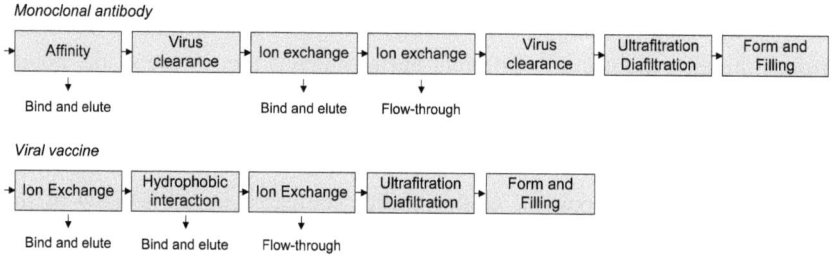

Figure 2: Typical downstream process schemes. Cultivation and cell removal is followed by downstream processing. Several chromatographic steps generate a product of high purity. Between the above steps, further procedures of concentration and buffer change are optional possible.

Due to the costs of the downstream processing related to the overall production process costs [1, 16, 33, 39-41] industry continues to search for approaches to skip process steps or using alternative technologies like ultracentrifugation. To reduce the development timelines and costs of production, manufacturers try to implement basic process platforms, for example, the use of a certain cell line with the same process steps for the production of different biopharmaceuticals [42-44]. The switch to disposable systems is currently a trend in the market place that aims to simplify process preparations and avoid expensive validation of equipment-cleaning procedures. Other possibilities are to improve the already well-known purification steps. General continuous downstream process steps are an improvement. For example, the simulated moving bed for chromatography has long been cited as being able to overcome limitations on the size of chromatography devices and the associated non-continuous processing of larger volumes [29, 45, 46].

3.2 Contaminants in downstream processing

Contaminants or impurities are unwanted substances derived from either the environment or the production process. In general, they are classified process- or product-related impurities. Product-related impurities are undesired variants of a product such as protein aggregates, aberrant glycoforms or oxidized forms [42, 47]. In contrast, process-related impurities mainly caused by the host cell organism, extractables of materials which come into contact with process media or components of the fermentation media. Added anti-foam agents or antibiotics will also need to be removed before final filling of the product.

The administration of contaminants, for example, subcutaneous or intravenous, can trigger the immune response of an organism. Further, it can act as a poison or reinforce the effects of a drug (adjuvant). The effect is short-term illnesses, but can also lead to chronic disease or death [35, 47, 48]. For this reason, biopharmaceutical manufacturers must ensure that their products are free of impurities. Limits were defined based on studies of the effect of various substances. Some substances will be removed below the detection limit, but generally accepted values are rarely found because this depends on the maximum administration of a certain drug to the patient. However, there are some suggested guidelines. The manufacturer has to demonstrate to the authorities that the impurities in question are within the suggested limits.

A large amount of impurities or contaminants are produced by the host cells itself. Cells contain e.g. components like proteins and lipids to enable cell functions and stabilize the cell structure. Some constituents, cytotoxins or degradation products are released already during the lifecycle, while others will be set free as a consequence of cell lysis. The major contaminant groups are the following:

Host cell proteins (HCP) are defined as proteinaceous cell constituents different from the target recombinant product [12, 49, 50]. Their various functions they are indispensable for the normal cell functions like e.g. growth and multiplication. Differences are in size, shape, physical and chemical properties such as the isoelectric point. This diversity often makes it necessary to use several purification mechanisms to achieve a sufficient clearance from the target molecule. The initial content of HCP can be up to several grams per litre. After a first affinity chromatographic step in the mAb processes, the concentrations were up to 10000 ppm [3].

Nucleic acids (Desoxyribonucleic and Ribonucleic acid) are essential molecules containing genetic information and are necessary for the synthesis of proteins. As a rule of thumb the amount should be 10 to 100 pg per dose for a given biopharmaceutical [51, 52]. In organisms DNA is present in the form of a double helix structure. Theses nucleic acids (NA) may appear either in small circular molecules or in chromosomes with lengths up to several millimetres. NAs are highly negatively charged because of the phosphate groups in the backbone of the molecule. Therefore, the method of choice is the removal with anion exchangers in the form of resins or membranes.

Viruses are infectious parasitic particles. The main components are viral DNA or RNA for replication together with structural proteins for e.g. the capsid. They are only able to reproduce with the aid of the metabolism of the infected cell. The risk of the presence of virus depends for some instance on the used cell line too. Particular features of viruses are the size of 5 to > 100 nm [53] that means tiny in comparison to other cell constituents but large to the soluble proteins like mAb. Furthermore, some viruses have an additional envelope outside to the capsid structure, a lipid bilayer with integrated proteins necessary for docking to and penetrating the cell surface [30, 36, 54, 55]. Different sources can be distinguished for viral contamination. Adventitious viruses are accidently introduced during the handling in process, e.g. by contaminated cell culture media, equipment and reagents, the operator or the latent infection of a cell line. Endogenous viruses originate from the host cell itself and are inherited as part of the genome [56].

Other contaminants, which can cause severe problems in humans like fever or anaphylactic shock syndrome, are endotoxins. They are lipopolysaccharides and can be released from the outer membrane of gram-negative bacteria [57, 58]. The effects are fever and inflammation. In severe cases, it can lead even to death by septic shock [59]. In general, endotoxins are omnipresent. For example, the concentration of endotoxins in drinking water is between 1 and 20 endotoxin units (EU) per millilitre. In gram negative bacteria cultivation e.g. *Escherichia coli* the endotoxin level can be high as 1.000.000 EU/ml. In animal cell cultures it is often less than 100 EU/ml [60, 61]. EU is the unit for the endotoxins which based on its biological activity. Thus, it can be vary between different sources. The FDA released a reference standard of endotoxins, which is currently determined with 10 EU per ng endotoxin [62]. Due to their negative effects, endotoxins must be removed to < 0.25 EU/ml for sterile water for injection [63].

Aggregates are incorrectly folded and/or associates of proteins and have to be removed from the correctly folded native proteins. Non correct folding of the protein structure can lead to hydrophobic amino acid residues being exposed to the surface of the protein. Upon that change in the conformation, interactions with other proteins can lead to aggregation to larger complexes. Strong aggregation of some proteins can cause protein precipitation, while others retain their solubility [9, 48]. Aggregates can be formed by chemical, thermal or other physical effects. Mechanical shear can destroy the structure of proteins [64-68].

An example is the bursting of bubbles in the fermentation reactor, leading to damaged proteins appearing in the foam [69, 70]. In certain buffers or at low pH, proteins are less stable and tend towards aggregation [71]. The amount of aggregates to be removed varies at different processes: in a mAb process it can be between 1 and 20 % of the target protein [3]. The proportion of e.g. mAb aggregates often increases with increasing amount of protein present [71, 72]. Furthermore, the process may be reversible. A change in buffer conditions helps to dissociate the aggregate complex into the monomer units [73].

Furthermore, during the downstream process different contaminating substances, called leachables, can be released from the equipment used. For example, fragments of the ligands of chromatographic resins are released in an affinity chromatographic step from a Protein A resin [28]. Leachables are suspected to promote the formation of aggregates [74]. Compared to other contaminants the levels are rather low [3]. But Protein A is a certain concern because of the multiple effects of that molecule to the human organism [75].

The evaluation of the effect on product purity and thus adverse reaction by different contaminants is complex. Process impurities often cause product-related ones by e.g. increasing the formation of product aggregates [22]. Proteases can degrade the target proteins [11]; contaminants often affect each other thus abolishing the predictability and prevent a generalization of the contaminant composition.

In addition to the removal of contaminants during downstream processing, a general goal is the prevention of their formation. A proper selection and correct operation of expression system or adding the amino acid arginine into the Protein A affinity chromatography step to reduce protein aggregation are only two examples described [50, 76].

Much attention during the purification of pharmaceutical products like mAb is thus paid to the detection of possible contaminants. Table 1 lists relevant analysis methods for the most important substance classes. The respective assays used in this thesis are described in the Appendix.

Table 1: Generic methods for contaminant detection and quantification.

Substance	Assay	Remarks
DNA	Picogreen, PCR[1]	Real-time PCR[1]
Host Cell Proteins	ELISA[2]	
Leached Protein A	ELISA[2]	
Aggregates	SEC[3]	Detection at 280 nm
Viruses	PCR[1], plaque assay	Model Virus ΦX174
Endotoxin	LAL[4]	Gel clot / chromogenic

[1] Polymerase chain reaction
[2] Enzyme linked Immunosorbent
[3] Size exclusion chromatography
[4] Limulus Amebocyte Lysate

3.3 Membrane adsorbers

In general membranes used in downstream process applications are a separating layer exhibiting sieving effects like e.g. virus or sterile filters. Microfiltration is defined at pore sizes rating from 0.1 – 15 µm, ultrafiltration with nominal molecular cut off values ranging from 2 – 1 Mio Dalton. This classification is somewhat arbitrary. The performance of membrane adsorbers is based on the same mechanisms exploited in liquid chromatography with resins. A mixture of substances is separated by different distribution between the solid stationary phase (here: membrane) and a mobile liquid phase. For membrane adsorbers, also known as membrane chromatography, ligands are attached to the large inner surface of a membrane matrix by e.g. grafting or coating processes [77, 78].

The membrane adsorbers, which were used in this thesis, are based on reinforced regenerated cellulose with a nominal pore size of 3 – 5 µm [79, 80]. An advantage of this membrane material is the low nonspecific binding for various biomolecules. This eliminates unwanted and difficult-to-control adsorptive mechanisms. During manufacturing a polymer chain structure is applied (grafting) to the entire inner pore surface or to partial regions of the basic membrane respective, e.g. polyacrylates with epoxy groups. These chains are forming the reactive layer for the attachment of various ligands. The chromatographic active layer in the pores typically has a thickness of less than 1 µm leading to a negligible diffusion limitation of mass transfer which is the most important feature of membrane adsorbers compared to commercially available resins on the basis of agarose e.g. Sepharose™ or other beaded polymers [17, 80-84]. An additional feature is ready-to-use i.e. out of the box, simple handling and operation and single use application. A feature of interest of adsorptive media is the binding capacity for the target. Although membrane adsorbers are planar structures, for a comparison the capacity is given not only as weight per membrane area but also in weight per volume. This originated from a historical perspective, since the amount of resins is given as a volume. Some ligands and the separation mechanism used for membrane adsorbers and resins are listed in Table 2.

Table 2: Chromatographic ligands of membrane adsorbers and interactions. In contrast to weak, strong ion exchanger binding properties are less sensitive to the pH. The ligands are present in dissociated form in broad pH range.

Ligand / functional group	Interaction
Sulphonic acid	Strong cation exchanger
Quaternary ammonium	Strong anion exchanger
Primary amine	Weak anion exchanger
Protein A	Affinity
Phenyl	Hydrophobic
Iminodiacetic acid	Metal chelate

In commercially available devices often a stack of several layers of flat-sheet or wound membranes are used to increase the built-in membrane volume [27, 81]. Applications are described where the process fluid is passed tangentially over both sides of the adsorbing membrane thus allowing the processing of particle-containing solutions [85].

In contrast to the open-porous structure of membrane adsorbers, column chromatography resins consist of a packed bed of porous chromatographically active beads. Thus, resins have a higher specific inner particle surface which leads to a higher binding capacity compared to the macroporous membrane-based adsorbers. Due to the high packing density and dead-end pores of the beads a disadvantage of resins is the molecular diffusion into and out of the pores of the beads. Here driving forces for the mass transfer are Brownian motion and a concentration gradient [86, 87]. Therefore, the time required for the molecules to reach a potential ligand is much higher than for membrane adsorbers where the mass transfer occurs mainly by convention rather than diffusion [77]. Thus, the binding kinetic of molecules is independent in a broad range from the residence time for membrane adsorbers, but not for resins. Furthermore, the packing density of a packed bed is high and causes a high flow resistance and hence a high pressure drop. From a certain bed height usually 10 – 15 cm the scale up is only possible by the enlargement of the column cross-section. Although the binding capacity of resins for small molecules is higher than for membrane adsorbers, larger molecules are less or unable to diffuse into the pores. Larger target molecules which bind to the outer region of the beads blocked the pores and prevented the entrance of other molecules into the interior of the bead [16, 41, 78, 80, 88]. This effect additionally is an advantage of membrane adsorbers for the adsorption of larger target molecules. A higher binding capacity for viruses is described [43].

Further alternatives to conventional chromatography in downstream processing are chemically modified hollow-fiber modules [89] or monoliths [90, 91]. These systems also have the goal to reduce the flow restrictions and limitations in capacity of resins. Moreover, like membrane adsorbers they improve the handling since no packing of a gel bed is necessary.

One of the most generic steps in contaminant removal is ion exchange chromatography. The separation step is based on the charge of molecules at different pH values. Depending on the pH, the ionized amino acid side chains of proteins influence the net charge [92, 93]. For example, the negatively charged groups predominate, thus the protein binds to a positively charged ion exchanger. The isoelectric point (pI) describes the pH where the net charge of the protein is zero. If the pH is below the pI, a protein is positively charged. A high pI of a monoclonal antibody favors the use of an anion exchanger in the neutral pH range because most contaminants have a negative charge (Figure 3).

Theoretical backgrounds

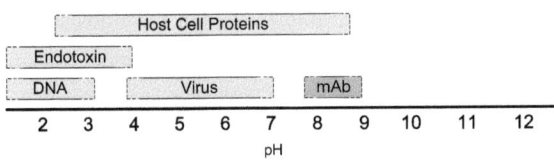

Figure 3: Charge of contaminants and antibodies. At a pH around 7 a polishing step using an anion exchanger is preferred in flow-through mode due to the negative charge of most contaminants and a positive charge of the target molecule (mAb). The distribution of contaminants reflects an ideal situation. Measurement of the isoelectric points of viruses revealed that a binding at neutral pH is also possible [93].

The strength of binding a protein to the ligand depends on the amount of charged chains and their accessibility. However, proteins and equally charged ions compete for the adsorption sites. The affinity of salt ions is generally higher than of proteins. This can be used to elute bound protein molecules from the ion exchanger. Furthermore, proteins with different electric charges can be separated by adjusting the ionic strength e.g. continuously by applying a salt gradient. Since salt ions present in process solutions can prevent the binding of target components need for dilution which leads to higher process volumes and requires high volume vessels on diafiltration to remove salt, the aim is to develop ion exchangers which also bind e.g. proteins at higher ion concentrations in the solution. These are referred as salt-tolerant ion exchangers [94, 95].

3.4 Bottlenecks in process development

Previous chapters provide an overview and illustrate the complexity of biopharmaceutical production. Nowadays there are expression systems with a challenging array of biological impurities and different chemical properties of the cell culture media. This is accompanied by a limited knowledge of contaminant composition and their mutual interactions or influences on the different process steps. Nevertheless, it is desirable to design the downstream process at an early stage. A robust process with an effective purification train accelerates the entry into the time-consuming clinical trials where effects and adverse reactions of the biopharmaceutical are investigated. Furthermore, regulatory requirements increase the demand on understanding the process steps thus developing drugs and processes which promote lowest health risks for the patient during administering the drug [96-98].

The development of a biopharmaceutical production always starts at small scales. After selecting a suitable modified microorganisms or cell line, the cultivation and the production of the target molecule are optimized. This is performed in the so-called upstream process. Furthermore, only small amounts of fermentation broth or cell culture medium is available for simultaneously establishing the purification strategy (downstream). Later in the development of the purification train it is often difficult or even impossible to change a purification step, because if a procedure is changed or replaced at a larger scale, it can have an impact on the product quality. This would unnecessarily boost costs and slow down progress due to additional regulatory effort and trials in clinical phases. Therefore, effective methods and small-scale devices are required. So it is obvious that new technologies are difficult to establish, because small-scale approaches are not available or insufficiently scalable.

There may be similar problems with development of purification technologies. For instance in chromatography more suitable ligands or matrices would be valuable and necessary. Performance influencing variables which can be manipulated are e.g. the pore size or ligand density of an ion exchanger membrane. To characterize the performance of new adsorptive media model proteins like bovine serum albumin are used [41, 95, 99]. However, model systems do not represent a real application [100]. The binding behaviour of target molecules from complex media is often completely different and interactions are sometimes unpredictable. A structural change of a membrane not only may have an impact on its binding capacity, but also on the selectivity [54]. Even more difficult to predict are chromatographic approaches with mixed mode adsorbents. Such media combine different adsorption mechanisms like ionic exchange and hydrophobic interaction [31, 101, 102].

The limited availability of real process media and a high number of influencing parameters accelerates the development of new small-scale purification screening approaches. To get rid of these development bottlenecks several techniques have been established. As a "toolbox" high-throughput screening (HTS) techniques are well known in the pharmaceutical and chemical industry, especially in the drug discovery [103]. In general high-throughput describes the parallelization of simultaneous running operations. It enables a systematic examination of a multiplicity of parameters or combinations, for example, the investigation of several additives on cell growth during fermentation. This includes a high statistical reliability and less consumption of raw materials. Important topics of HTS are experimental design, automation and data analysis.

In the last decade high-throughput downstream screening approaches have emerged. The field of application mainly addresses chromatographic steps in the downstream development with resins. In 2005 one of the first resin-based 96-well product was described [104]. It consisted of a combination of several small-scale columns and was designed for a parallel column chromatography with a liquid-handling robotic workstation operated similar to large-scale columns, wherein the mobile phase flows through the packed bed continuously [100]. A second approach emerged with resin filled 96-well plates for screening. In 2010 the screening of process conditions for a Protein A affinity chromatography capturing step was described. The study showed the approach for the optimization of buffers for protein binding, washing and elution regarding contaminant removal and product recovery. In contrast to small resin columns the use of 96-well plates is a batch adsorption process [106]. The small cavities with a few µl of resin are filled successively with the solutions used in a chromatographic operation (Figure 4).

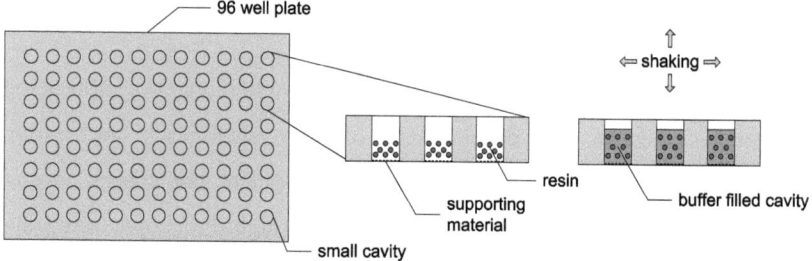

Figure 4: Chromatography resin screening. 96-well plate for resin small-scale screening approach. Small cavities are prefilled with 2-50 µl resin. After filling with liquid the plates will be shaken. A centrifuge or pressure gradient is used to remove the liquid. Resin beads are retained by the supporting material at the bottom of each well.

To mimic the flow in the column the liquids are shaken until equilibrium. The resin is placed on a supporting material and after the incubation the supernatant liquids are after centrifuged and aspirated for analysis [107, 108]. The batch experiments enable the mimicking of a whole chromatographic process (Figure 5). As mentioned, residence time is fundamentally important parameter for the operation of resins. Its determination is a basic application when screening chromatographic resins. Results show a shifting of the separation optimum by changing the residence time.

Figure 5: Chromatographic steps. Depending on the mode of operation only the necessary chromatographic steps are performed. Conditioning and equilibration prepare the adsorbent. Manufacturing-related impurities are removed and the equilibrium condition between the solid and mobile phase are reached. The load is the application of the solution containing target substances and contaminants to the column. In bind and elute mode additional wash steps shall remove contaminants. Regeneration is added if the adsorber material is to be used again.

High-throughput applications need technologies for a sufficient sensitive analysis of samples. Due to the standardized formats (96-well) and automation there are only limited restrictions. The investigation of HCP or antibody purification with Protein A resins using high-throughput approaches are described [109]. Furthermore, a platform technology including the automation of chromatographic steps and analysis of samples for the screening of chromatographic resins is demonstrated in downstream processes [106].

In terms of membrane chromatography a simple calculation illustrates the demand on material required for the investigation of a polishing step. For contaminant removal at the end of a process chain only a small amount of impurities in relation to the target molecule are remaining. Thus, only a small amount of adsorptive media is required. Because of high possible flow rates membrane adsorbers are ideally suited and an oversizing is not necessary like with resins. But scale down of these processes conditions cause a problem. Due to the high binding capacity for the residual low concentrated contaminants the total load of the target protein per adsorber volume is high. Studies showed that up to 15 kg per litre of membrane can be processed [110, 111]. For small scales this means even for a membrane volume of 1 ml up to 15 g mAb is necessary to reflect a process condition. This highlights the need for small-scale approaches in membrane chromatography. Figure 6 summarizes the use of high-throughput screening. It contains the application and membrane adsorber development. In addition to the optimization of the parameters for applications, screening is a tool to investigate process strategies and robustness. In membrane modification and manufacturing, the use of different model contaminants or proteins offers a rapid evaluation of the parameters influencing the manufacturing process.

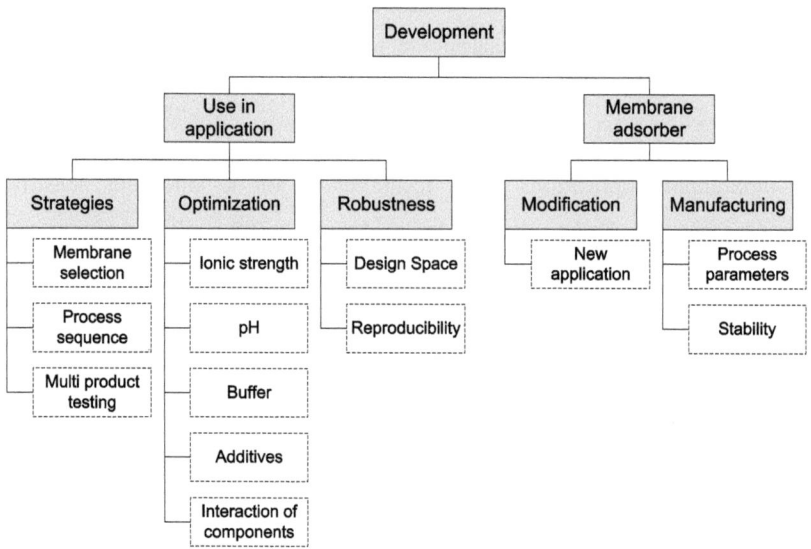

Figure 6: Use of screening in membrane chromatography. The chart presents an overview of potential use in the application- and membrane adsorber development. Strategies: the selection of chemistry and step position in the purification process. Optimization: Investigation of the influence of buffer composition and type of proteins. Robustness: Exploration of the design space to assess the effect of process variations on the chromatographic step and its robustness. Modification: The use of model systems triggers the evaluation of new ligands and applications. Manufacturing: Effects of changes in process parameters like temperature or chemicals on the stability of the membrane surface (homogeneity).

Several small-scale approaches for high-throughput screening using membrane adsorbers were published. In Table 3 96-well high-throughput applications for resins and membrane adsorbers are summarized. Furthermore, HTS applications applied in bioprocess development are reviewed in [108].

Table 3: High-throughput applications in resins and membrane based chromatography.

Published	Screening media/device	Application
2005 [112]	Resin 100 µl/well, 96 well plate	Purification of recombinant proteins
2006 [113]	Membrane adsorber 1.6 cm²/well (44 µl), 8-strip 96 well plate	Isolation of model proteins and polyhistidine-tagged proteins
2007 [109]	Resin 200 µl/well, 96 well mini columns	Removal of HCP from mAb
2008 [105]	Resin 50-100 µl/well, 96 well plate	Removal of high molecular weight product-related impurities from mAb
2008 [115]	Resin 100 µl/well, 96 well plate	Removal of aggregates from mAb
2008 [116]	Resin 50 µl/well, 96 well plate	Removal of HCP from mAb
2008 [117]	Resin 2 µl/well, 96 well plate	Binding of polyclonal human immunoglobulin and amyloglucosidase
2010 [118]	Resin 25 µl/well, 96 well plate	Removal of HCP and aggregates for a polishing step in mAb production

4 Implementation of a 96-well HTPD platform for the characterization of membrane adsorbers

4.1 Device and screening platform

The first chapter of the practical part of this thesis deals with the development of a suitable screening approach for membrane adsorbers. This includes the construction of a device which enables the operation with flat-sheet membrane layers. The properties of the device are tailored to work with contaminants as mentioned in Chapter 3.2 and fulfil the screening format requirements. To accelerate the screening process the device is adapted to an automated liquid handling platform. Regarding the chromatography steps the main performance characteristics are explained and examined.

4.1.1 Development of a reusable 96-well device

Essential features for the device are dictated by the limited availability of raw material and sample size. In small-scale fermentation only small amounts of process media are available, also in membrane development processes chemical-modified samples are available only in small quantities. Therefore, to obtain as much of information about a specific modification of a membrane as possible a simple approach is required. Furthermore, a miniaturization should enable a parallelization to reduce the required time for generating characteristic data in membrane chromatography. The small-scale approaches for membranes described or commercially available are based on multi-well devices with separate membrane material i.e. the incorporation of discrete membrane pieces into each single well of the plate which causes problems with sealing and increases complexity of assembly like used in [113, 114]. The use of well plates with consist of chromatographic resins is time-consuming because of required residence time to achieve equilibrium [117]. To circumvent these obstacles a holder was developed which enables the use of single continuous flat-sheet membrane adsorber layers, which forms discrete wells. A sealing prevent a cross-talk caused by lateral diffusion between individual wells.

Figure 7 and 8 show the holder construction whose dimension are adapted to the standard format of consumables like 96-well plates [119] or UV-transparent plates suitable for analysis. The holder is compatible with standardized robotic liquid-handling systems.

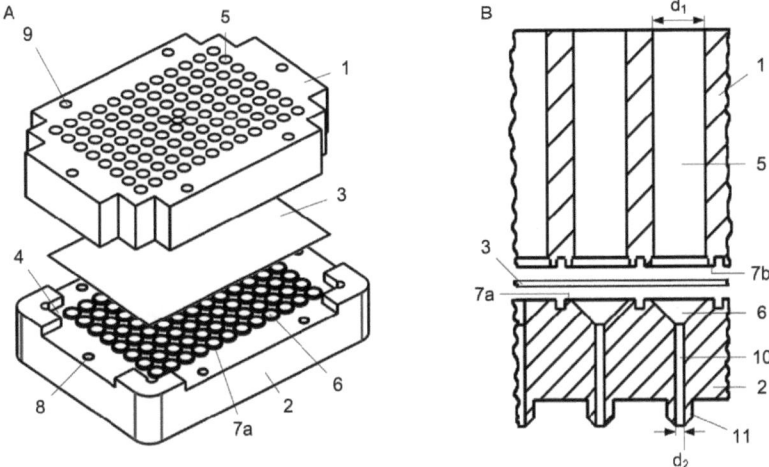

Figure 7: Homemade 96-well membrane holder construction. A: Exploited view of the device with membrane adsorber sheet; B: cross-section of the device; (1) top part, (2) bottom part, (3) stack or single sheet of membrane, (4) membrane stack receiving surface, (5) cavity for applying samples, (6) permeate outlet funnel, (7a) sealing surface bottom part, (7b) sealing surface top part, (8) threaded bore, (9) bore for screwing bolt, (10) capillary, (11) drip nozzle, (d_1) diameter of cavity for filling 6.6 mm, (d_2) diameter of capillary 1 mm

The multi-well holder consists of a top and bottom part; between the parts a membrane adsorber stack or single sheet can be fixed. The top part of the holder has 96 cavities with a respective holding volume of 500 µl. Together with the fitting bottom part filtration chambers (wells) are formed. A transverse section of a well is 0.34 cm². The total amount of adsorbent per well can be set by the number of membrane layers. The sealing is performed by compression of the sealing surfaces. The form of the bottom part (permeate side) ensures a uniform flow distribution and a separate collection of permeates into the discrete wells. The bores form a funnel and then a capillary. This ensures an effective prevention of cross-talk between the wells. So the effective filtration area of each well is clamped in a contactless manner. The small diameter of the capillary ensures complete flushing and/or emptying of the fluid path. Driving forces for the permeation is a pressure gradient by vacuum or gravity.

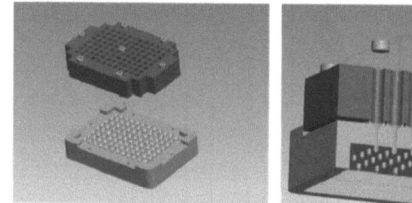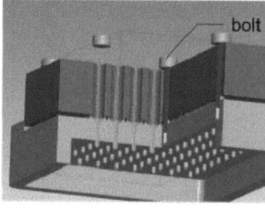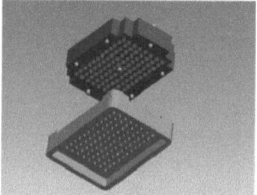

Figure 8: 3D CAD construction of the 96-well membrane holder. By means of 7 threaded screws the membrane stack is compressed between the bottom and top part of the device. A uniform sealing is achieved by a criss-cross screwing with torque of 4 Nm.

The non-destructive assembly of the device allows reuse. The holder is made from aluminium and can be heated up to 200 °C. It can be depyrogenated if it is necessary. Metal can bind molecules to their surface [65]. The holder was anodized to smooth the surface or avoid oxidation by caustics or acids.

Since hydrophilic membranes are used, diffusion of aqueous solutions occurred due to capillary action leading to detrimental cross-talk which was caused due to the installation of dry membrane layers into the holder. To prevent this, the membranes were wetted with buffer before installation. In the following, the cross-contamination (cross-talk) between the wells was studied. In the first method dyes were used to stain the different membrane types under investigation. Due to the ionic interaction dyes like Ponceau S, Brilliant Black or

Methylene blue were used. As the device accommodates various layers, the holder was equipped with a different number of layers. Each well was then filled with a staining solution. To exclude the influence of the residence time, the filtration started 15 min after filling each well. This selected time was higher than for the filtration where the residence time per well is less than 1 min. Figure 9 demonstrates the absence of lateral diffusion i.e. no cross-talk. For easier addressing the loading of liquids, the 96-well holder was generally subdivided into an 8 x 12 matrix (8 rows and 12 columns) for all subsequent studies.

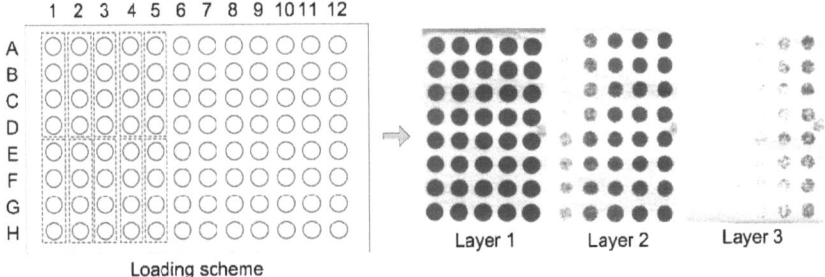

Figure 9: Prove of absence of lateral diffusion. The membrane holder was equipped with a stack of 3 layers of anion exchange membrane (quaternary ammonium). Different volumes of a solution of Ponceau S (43.5 mg/L) in 20 mM Tris/HCl buffer (pH 7.4) were loaded into 40 wells. Left: The coordinates of the loading scheme describe the pipette volumes: A1-D1 400 µL, E1-H1 800 µL, E1-H1 1200 µL, E1-H1 1600 µL, E1-H1 2000 µL, E1-H1 2400 µL, E1-H1 2800 µL, E1-H1 3200 µL, E1-H1 3600 µL, E1-H1 4000 µL. The remaining well were filled with 500 µl water. The increasing load volume was chosen to visualize the breakthrough. After the holding time and filtration the membrane layers were removed and inspected. Layer 1 was on top.

These previous results were no proof that non-binding small molecules like buffer constituents will as well not show lateral diffusion. Since phosphate ions are such a common component, a detection of this component was checked to exclude cross-contamination. The phosphate detection assay of Fiske and

Implementation of a 96-well HTPD platform

Subbarow [120] was employed, which is explained in the Appendix. If phosphate ions are present in a sample there will be a dark blue color change. Thus, different wells were filled randomly with a sodium phosphate solution, the remaining with water.

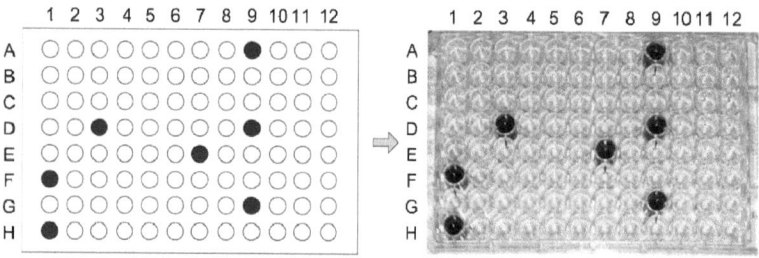

Figure 10: Trial to detect cross-contamination using phosphate ions. Each well was filled with 500 µL water or sodium phosphate solution (0.16 mg/ml). The filtration was performed using a vacuum. Samples of the permeate were mixed with the detection reagent. The detection limit of the assay was 2 µg/ml. The blue coloration of the liquids and measurement of the absorbance at 820 nm indicate the presence of phosphate using the Tecan Safire Plate Reader. Blank was 564 ±58; sodium phosphate solution was 34948 ±245 arbitrary units.

After 15 min vacuum of 350 mbar was applied, the filtrates were collected and the presence of phosphate was checked. In the samples of neighbouring wells filled with phosphate buffer, the concentration of phosphate was below the detection limit (see Figure 10). The results presented were performed with three layers of membrane.

4.1.2 Automation

Aim of a high-throughput approach should be the processing of a plurality of samples with high statistical reliability. For the device described in the previous chapter an automated operation scheme was developed. The result of the development was a high-throughput downstream screening platform, suitable for an automated characterisation of membrane adsorbers. The structure and working principle of the platform is described in the following.

The workstation was an automated liquid-handling platform coined Lissy 2002 from Zinsser Analytic [121]. Figure 11 shows the schematic top view of the system. For sake of clarity the moving needle arm or the gripper arm are emitted. The robot operates without a change of pipette tips: 8 pipetting needles are used and can be cleaned in a washing station. With these needles it is able to mix buffers and protein solution at small scale. A data import was programmed. The buffer compositions can be keyed into an Excel template, regarding species, conductivity, pH and concentration. Then the values are transferred to the control software of the liquid-handling platform and calculated by previously stored calibration curves. The stock solutions prepared by hand and are placed into the storage tanks. It is possible to mix up to 80 buffers in the tubes see Figure 11 (B). The filtration station to accommodate the 96-well membrane holder is located in the centre (see Figure 11). The commercially available vacuum station consists of a reception for depth-well plates to collect different fraction of the filtrate and a port to connect a vacuum pump (D). The membrane holder is placed on the top of the station. Furthermore, several stackers are used for depositing several depth-well plates and 96-well UV plates (F). A gripper arm is able to transport these consumables to different position (not shown). This includes the remove of the membrane holder or the transport of the 96-well UV plates to a connected 96-well plate reader.

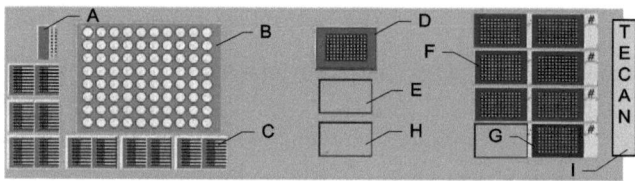

Figure 11: The liquid-handling system. Schematic top view of the automated platform. (A) needle wash station, (B) falcon tubes for buffer mixing, (C) reservoirs for buffer stock solutions, (D) vacuum station, (E) deposition rack, (F) depth well collection plates, (G) 96-well UV transparent plates, (H) sample station, (I) 96-well plate reader. A full processing cycle including the preparation of 80 buffers lasts 1-4 h (flow-through/bind and elute).

The operation of the filtration station was modified to enable a processing with vacuum or pressure application. Before each filtration step a partial vacuum was generated by the vacuum pump. Filtration started when a defined pressure has been reached in the reservoir. The level of the applied vacuum influences the filtration rate. A valve was used to regulate the pressure of the compressed air during the filtration. A slight backpressure ≥ 5 mbar inside the filtration station was applied to prevent premature flow of the liquid by gravity force during the filling of each well.

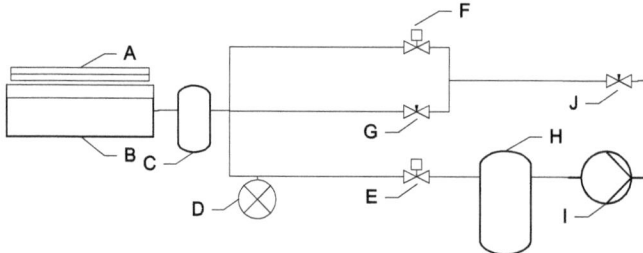

Figure 12: Circuit diagram vacuum station. (A) 96-well membrane holder, (B) vacuum device with 96-well collection plates each well with 2 ml capacity, (C) overflow protection, (D) pressure gauge, (E) pneumatic valve for vacuum application, (G) valve for pressure regulation, (F) pneumatic pressure valve, (H) vacuum reservoir, (I) pump, (J) compressed air supply.

Implementation of a 96-well HTPD platform

The sequences of the movement of the both arms described above can be programmed into an input mask displayed on the controlling computer. The program then performs the automatic execution of the chromatographic steps. A program structure is schematically summarized in Figure 13.

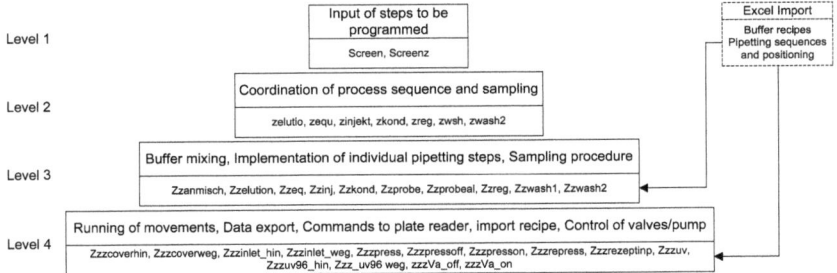

Figure 13: Screening program architecture. Through the user interface the individual chromatographic steps can be selected and the analysis steps controlled. The program structure is subdivided into 4 levels. Level 1 consists of the user input. The sequence of steps and the assignment of loading schemes and buffer mixing recipes are implemented using Excel templates. Level 2 coordinates the final program sequence. Levels 3 and 4 are sending the executing commands to the hardware. The individual programmed subroutines are named in the lower part of each level.

The liquid-handling platform software 'ZA runner' then executes the programmed pipetting steps.

4.1.3 Liquid-handling

The critical factors using automated liquid-handling systems (LHS) are uptake (i.e. suction), transport, dispensing (delivering) of different liquids and cleaning of the needles [108]. Because air is compressible and its presence in the tubing would lead to erroneous results and the needles have to be cleaned after every pipetting step, all tubing and tips are always filled with water. Thus, during the uptake of liquids there would be a mixing of the sample with water. To prevent this, before sampling a certain amount of buffer and a small air bubble is picked up. This will remove residues of the liquid and create a barrier between the water and the sample.

Furthermore, the speed of uptake and dispensing has to be carefully controlled. A too high speed of suction will lead to a destruction of the air bubble barrier. Additionally, during the sample preparation the pipetting of the protein solution is performed slower than for the other buffer components and is dispended directly in the sample liquid to avoid protein aggregation. The sampling of mixed buffers or collected fractions also is performed slower too. After each contact with a protein solution, the tips were cleaned by three successive steps of 5 ml water or detergent. The design of the wash station allows the rinsing of the outer surface of the needles by immersing in flushing liquid. The different pipetting parameters determined for the experiments are summarized in Table 3.

Implementation of a 96-well HTPD platform

Table 4: Liquid-handling Parameter (LHP). The pipetting operations are divided into basic groups namely mixing in Falcon tubes, pipetting to the 96-well screening device and sampling. Level detection controls the movement of a tip down to the liquid surface before uptake or dispensing liquid, level tracking the associated movement. This is for keeping the exterior surface of the tip clean. Furthermore, it prevents the denaturation of protein due to shear forces (foaming). Then, the tips move to the predetermined position. A short delay after dispensing reduced drops, after uptake the fluid enters completely the tip. After the buffer addition into the respective receptacle the content is mixed by pipetting water at a high flow rate with the tips dipping into the liquid. The velocity affects also the stability of the protein and the accuracy of that pipetting step. A wash after each step is only necessary when the liquid or buffer is changed. During sampling it reduces deviations.

LHP/ Step	Variable	Mixing buffer	Mixing Protein	Screening pipetting	Sampling
Uptake	Immersion depth of pipetting tip [mm]	7	7	7	2
	Liquid/air bubble before uptake [µl]	20/20	20/20	20/20	None
	Uptake velocity [µl/s]	500	100	100	100
	Level tracking / Liquid detection	None	Yes / yes	Yes / yes	None
	Delay after receiving air/liquid [s]	0.2/0.5	0.2/1	0.2/0.5	0.2/0.5
Dispensing	Immersion depth of pipetting tip [mm]	7	7	0	0
	Dispensing velocity [µl/s]	500	150	150	150
	Level tracking / Liquid detection	Yes / yes	Yes / yes	Yes / yes	Yes / no
	Delay after dispensing	0.2	0.5	3	0.5
	Forced wash after each step	Yes	Yes	None	Yes

The accuracy of the pipetting depends on the diameter of the needle and tubing and the minimum step size of the piston pump motor. The manufacturer of the LHS specifies a standard deviation of less than 1 % per tip and < 3 % between all tips. The optimum working volume is 10 % of the syringe volume: this corresponds to 600 µl. In order to reduce the variations in the marginal boundary area, a volume correction factor (Volcor) was introduced for each pipetting tip. With this the LHS software adjusted the number of motor steps to fit the actual amount of pipette volume to theoretical value. The adjustment was calculated by comparing the set point (set$_i$) and actual value (actual$_i$). The new volume correction factor was calculated by n measurements of the weight of a specific pipetting volume. The density was assumed to be 1 g/ml. The formula is given in (4.1)

$$Volcor_{new} = 1 - \frac{\sum_{i=1}^{i=n}\frac{actual_i}{set_i}}{n} + Volcor_{old}$$

(4.1)

The factor was determined for 50, 100, 300, 500, 800, and 2000 µl. For 5000 µl no factor could be evaluated due to limited motor steps. The selection was verified by typical pipetting volumes like sampling or filling the wells. Motor steps for a given volume between the values were calculated by linear interpolation.

To determine the accuracy of pipetting steps in the area of typically used volumes during the operation, the actual amount of dispensed water of each needle was determined. The evaluation of pipetting errors later allows a differentiation between the deviation caused by the liquid-handling system, plate reader and membrane. Furthermore, the error accumulation could be estimated. Using the previous determined volume correction factor Table 10 in the Appendix (8.3.1) summarizes the characteristic values of the pipetting steps, i.e. the variation during the repetition of uptake and dispensing of each tip. As expected the volume correction factor and the variation increased for smaller volumes.

The next issue of the liquid-handling procedure the filling of the wells was established. For screening different possible applications, a method was developed which allows each chromatographic step of a well to be defined independently. The described program structure was designed for use an Excel template sheet to enter the allocations of different chromatographic buffers and wells on the LHS, i.e. the 96-wells, the position of the 80 falcon tubes for the required buffer position and the volumes to be loaded. Furthermore, a calculation sheet for different buffer systems to adjust pH, conductivity and molarities of the buffers was established. The calculations are explained in the Appendix.

4.2 Screening of membrane adsorbers

After the device and the working platform had been validated regarding installation, operational and performance qualification, the membrane adsorber screening platform had to be evaluated regarding the performance and influencing factors. The characterizations were done with stacks of three membrane adsorber layers. Thus, the membrane area per well was 1 cm^2. This corresponds to a bed volume of 28 µL. This size is based on the typical binding capacity of the membranes adsorbers for model proteins [95, 122] and an appropriate loading volume per step. For avoidance of error accumulations the loading steps were kept small, i.e. the minimum volume per step was 400 µl because the necessary sample volume for the plate reader was 300 µL. Furthermore, due to the possible inhomogeneities of membrane adsorbers resulting in uneven ligand density three layers of membrane were used.

4.2.1 Heterogeneity of membrane adsorbers on a small-scale format

For a successful use of the 96-well membrane holder the deviation between individual wells must be small performing a chromatographic step under same conditions. An important factor is the deviation of membrane regarding ligand density and flow. Hence a trial was performed in which each individual well was run at the same chromatographic conditions. Bovine serum albumin and an anion exchanger with quaternary ammonium ligand were used for this model process. After conditioning and equilibration the same amount of protein solution was loaded into each well. Then a washing step and the elution were performed. The pressure was reduced to 300 mbar during the filtration steps. All fractions were analysed with the plate reader at 280 nm. After the measurement of the flow-through fraction of each well and calculating the binding capacity by subtracting from the added amount of protein the variation coefficient υ, the ratio of standard deviation σ to the arithmetic average μ, was determined, where n is the number of samples and X_i the measured value. The evaluation includes the variations caused by pipetting steps and analysis.

$$v = \frac{\sigma}{\mu} = \frac{\sqrt{\frac{1}{n-1}\sum_{i=1}^{i=n}(X_i - \frac{1}{n}\sum_{i=1}^{i=n}X_i)^2}}{\frac{1}{n}\sum_{i=1}^{i=n}X_i} \times 100 \qquad (4.2)$$

The calculation of the binding capacity BC_{FT} per membrane area was done for each of the 96-wells by

$$BC_{FT} = \frac{(c_{Load} - c_{Flow-Through}) \times V_{Load}}{A} \qquad (4.3)$$

Where c_{Load} is the concentration of the loading solution, $c_{Flow-Through}$ the concentration (mg/ml) of the sample, V_{Load} the loaded volume (ml) and A the membrane area (cm²). It is assumed that the volume of the flow-through fraction corresponds to V_{Load}. Furthermore, the binding capacity BC_{El} was determined by analyzing the elution fractions, whereas $c_{Elution}$ is the concentration of the eluted protein (mg/ml), $V_{Elution}$ is the volume (ml) and A is the area of membrane (cm²).

$$BC_{El} = \frac{c_{Elution} \times V_{Elution}}{A} \qquad (4.4)$$

Addition of the measurement of the wash fractions the recovery R was calculated by

$$R = \frac{c_{Flow-Through} \times V_{Load} + c_{Wash} \times V_{Wasch} + c_{Elution} \times V_{Elution}}{c_{Load} \times V_{Load}} \times 100 \qquad (4.5)$$

Where c_{Wash} and V_{Wash} are the amount and volume of protein in the washing. Table 5 summarized the results. The similar values of BC_{FT} and BC_{El} indicate a nearly complete elution of the bounded protein.

Table 5: Evaluation of the variation of the measurement. Characteristic values using the 96-well membrane holder. 3 layers of an anion exchanger membrane with quaternary ammonium as ligand were used (1.03 cm² per well). After equilibration with 0.5 ml binding buffer (20 mM Tris/HCl pH 7.4), 2.6 mg/cm² of BSA were added in 2 steps of 0.5 ml each. The membrane was washed (0.5 ml) and protein was eluted by applying 1 ml of 1 M sodium chloride in binding buffer for determining the recovery rate. The amount of protein in the fraction was 0.07 mg (υ=24 %).

	µ	σ	υ
BC_{FT}	1.10 mg/cm²	0.05 mg/cm²	4.7 %
BC_{El}	1.01 mg/cm²	0.07 mg/cm²	6.5 %
R	100.6 %	3.2 %	n. a.

To estimate the contribution of the membrane and 96-well plate holder to the variation coefficient, both pipetting errors and deviation of the plate reader must be included. Parallel preparation of 8 protein solutions as used for the loading experiments the protein concentration 2.66 mg/ml ±0.065 (2.4 %). The variance included mixing pipetting steps, sample pipetting and the deviation of the plate reader. Furthermore, the two loading steps, sampling and analyzing together with the membrane increased the variation coefficient to 4.7 %.

4.2.2 Influence of the flow rate on the binding performance

It has been shown that the binding capacity of membrane adsorbers with quaternary ammonium IEX1 is independent from the residence time within a broad operational window. In contrast to larger-scale applications the flow rate using the screening device described here is higher, equalling to the residence time. In addition, during the experiments no uniform discharge of all wells was observed. Although the same pressure gradient was applied to all filtration chambers of the device during the filtration, different flow rates occur mainly due to the structure of the membrane or different viscosities of the filtration media.

To determine the average flow rates of all wells at various differential pressures three layers of membrane were assembled into the membrane holder. To each well 500 µl buffer were applied. The choice of buffer conditions was chosen appropriate to the application. By the addition of sodium chloride the conductivity was increased for trials with the salt-tolerant anion exchanger IEX2. Several reduced pressure values down to -350 mbar were adjusted. The weight of flow-through in the depth-well collection plate was measured after filtration within specified times. The weight increase corresponds to the average flow rate. $\dot{V}$ is the flow rate, m the weight of the empty and m_0 the filled collection plate and t the filtration time.

$$\dot{V} = \frac{m - m_0}{t} \times \frac{1}{96}$$
(4.6)

To compare the values with standard processes the residence time τ is represented by the inverse value of membrane volume per minute ($Flow_{MV/min}$), where d_M is the average thickness of the membrane and A the total assembled membrane area.

$$Flow_{MV/min} = \frac{\dot{V}}{A \times d_M} \quad (4.7)$$

$$\tau = \frac{1}{Flow_{MV/min}} \quad (4.8)$$

The flow rate depends on the type of membrane (Figure 14). Due to the matrix structure and the higher ligand density, the flows using the salt-tolerant anion exchanger are slower [122].

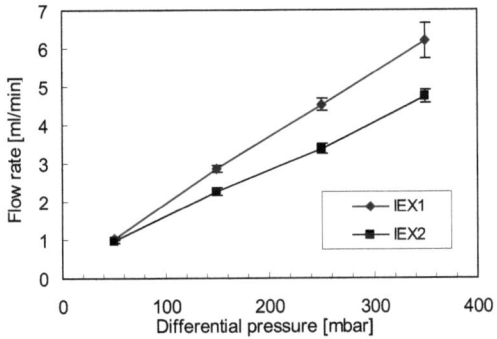

Figure 14: Flow rate depending on differential pressure. The flow rate was determined using the buffers 20 mM Tris/HCl pH 7.4 for IEX1, for IEX2 150 mM NaCl 20 mM Tris/HCl pH 7.4. The ligand of the anion exchanger IEX1 was quaternary ammonium, for IXE2 it was polyallylamine. For each pressure value n=3 measurements were performed. With an average thickness of 275 μm the flow rate at 350 mbar is 220 MV/min for IEX1 and 168 MV/min for IEX2.

Compared to larger membrane adsorber devices, where the maximum operating flow rate typically is < 30 MV/min [27], the residence time is increased by a factor of 7 (IEX1) and 6 (IEX2) using the 96-well filtration devices at -350 mbar.

An additional approach was used to estimate the variation of the flow rate between several wells. 800 µl of 2.7 mg/ml BSA were applied to each well of the collection plates. 500 µl buffer was filtered through a stack of three membrane layers at 175 mbar differential pressure. The filtration process was stopped after 5 seconds.

By analyzing the protein concentration in the flow-through fraction the calculation of the flow rate of individual wells was done with the formula

$$\dot{V}_i = \left(\frac{c_0 \times V_0}{c_i} - V_0 \right) \times \frac{1}{t}$$

(4.9)

Where $\dot{V}_i$ is the flow rate of an individual well, c_0 the concentration of the BSA solution applied to the collection plate and c_i the final concentration after filtration and t the filtration time. The coefficient of variation between individual wells was 50 %.

To assess the effect of the residence time on the binding performance model systems were chosen to mimic several contaminants. The first test system was BSA. For simplification it was assumed that the average residence time of the test solutions were equal to the buffer used. This was assumed to all further investigations studies described in the following.

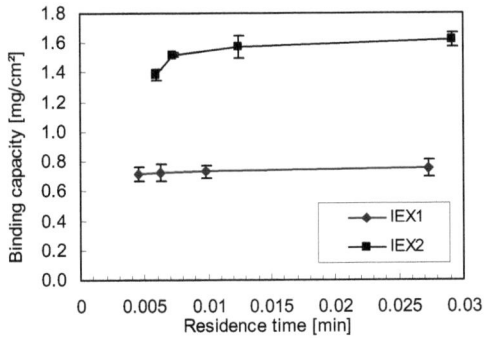

Figure 15: Binding of protein depending on residence time. 3 layers of IEX1 or IEX2 were used (1.02 cm² per Well). Buffer for IEX1 was 20 mM Tris/HCl pH 7.4, for IEX2 150 mM NaCl 20 mM Tris/HCl pH 7.4. After equilibration with 2.0 ml buffer per well, for IEX1 1.0 mg/cm² of BSA were added in 2 steps of 0.5 ml, for IEX2 2.3 mg/cm². Binding capacities were calculated for n = 12 wells for each residence time tested.

Figure 15 showed that with increasing flow rate the binding capacity was reduced by 5 % for IEX1 and 14 % for IEX2. These effects could be explained in part by the structure of the membranes. In contrast to IEX1 the salt tolerant IEX2 consist of "ultra pores", regions of the membrane with a dense structure [122, 123], in order to achieve a higher ligand density thus increasing the binding capacity. The loaded amount of BSA per membrane volume was chosen according to the expected binding capacity [95, 122], thus addressing all binding sites. The assumption is a reduced penetration into the "ultra pores" of IEX2 due to the short residence time, especially for larger molecules.

It was assumed that the behaviour was significantly higher for target substances with a high degree of contaminants needed to be removed (e.g. ≥ 99.9 %). Examples are viruses and endotoxins [33, 80, 84, 124-126]. Since working with pathogen viruses was not possible due to safety issues, phages were used instead as a model virus. The here used phage ΦX174 is described in Chapter 5.1. The concentration of infective virus particles is given in plaque-forming units per volume (PFU/ml). The endotoxin solution was prepared by diluting a commercial available concentrate. Both detection assays are explained in the Appendix. In downstream processing the removal of contaminants can be expressed as Log Reduction Value (LRV). The value describes the ratio of the concentration of a molecule in the initial solution c_0 compared to the filtrate c_i.

$$LRV = \log\left(\frac{c_0}{c_i}\right) \quad (4.10)$$

Figure 16 and 17 illustrate the influence of high flow rates on the removal of bacteriophages and endotoxin. It confirms the influence of residence time on binding of the virus particles. The same was true for the endotoxins which have phosphate as charged groups instead of charged amino acids in the case of ΦX174. The amount of phages or endotoxin in load was much lower compared to the maximum binding capacity of the membrane adsorber [88, 89, 122].

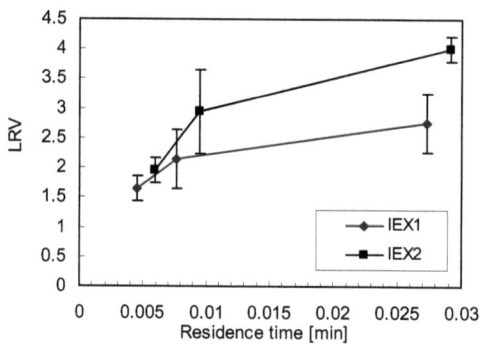

Figure 16: Binding of phages depending on residence time. 3 layers of IEX1 or IEX2 were used (1.02 cm² per well). Buffer for IEX1 was 20 mM Tris/HCl pH 7.4, for IEX2 150 mM NaCl 20 mM Tris/HCl pH 7.4. After equilibration with 2.0 ml buffer per well, 1.0 ml Bacteriophage ΦX174 spike solution was added in two portions. LRVs were calculated for n = 3 wells for each residence time. The initial infectivity titer was 1.2×10^7 PFU/ml.

Implementation of a 96-well HTPD platform

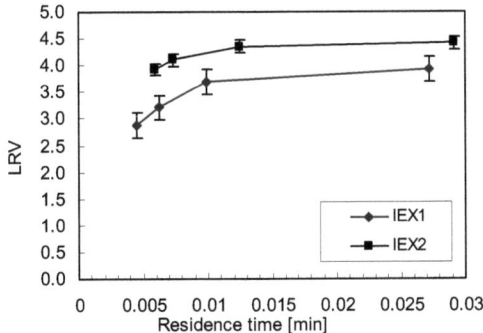

Figure 17: Binding of endotoxin depending on residence time. 3 layers of IEX1 or IEX2 were used (1.02 cm² per well). Buffer for IEX1 was 20 mM Tris/HCl pH 7.4, for IEX2 150 mM NaCl 20 mM Tris/HCl pH 7.4. After equilibration with 2.0 ml buffer per well, 0.5 ml buffer spiked with endotoxin was added. LRVs were calculated for n = 4 wells for each residence time. The endotoxin concentration c_0 was 500 EU/ml.

Although it has been shown that there are no diffusion limitations of mass transfer for membrane chromatography, two effects could explain the influence of short residence times on the binding as shown in Figure 15 – 17. The first is deterioration in the flow distribution of the 96-well device. A higher flow rate could lead to the formation of a flow profile in the center of each well. Thus, the throughput would be smaller at the edges of each well. Fewer binding sites would be available and the amount of non-bound molecules would increase. This assumption, however, was less favored due to the uniform staining of the membranes using Ponceau S shown in Chapter 4.1.1. An alternative effect could be the adsorption of gas as described in [127]. The binding mechanism could apply for proteins too. If a protein encounters a binding site, its energy is transferred to the surface of the adsorber or the mobile phase. A part of this energy also depends on the flow rate (kinetic energy). In the case of membrane chromatography, the binding interaction between ligand and molecules must be strong enough to overcome these forces and 'slow down' the molecule. The binding of a molecule can be induced by physical and chemical adsorption

[128]. The chemical adsorption is a covalent bond with a small distance between the adsorber surface and molecule. For the reversible binding of proteins to a membrane, physical adsorption takes place, e.g. the van der Waals forces. These are characterized by broad reach and weak binding forces. The rate of adsorption depends on the time required to transfer the energy to the adsorber (adsorption probabilities). If the energy cannot be transferred, the molecule bounces back into the mobile phase. The probability increases with fewer available binding sites. This was also assumed for the earlier breakthrough of molecules, even when sufficient ligands were available, because of the short residence time using 96-well plates.

The results address the effect of residence time in these small scale formats, which is much higher than for the larger membrane adsorber devices. To minimize this effect a slight backpressure was used during the filling of each well. So the filtration process could be started simultaneously. Another possible operation was the use of a permanent vacuum during filling of the wells. However, it appeared that a permanent vacuum caused a shift in performance even when using the same chromatographic conditions for all wells. An explanation could be that the membranes in the individual wells fall dry with time. This changes the permeability for air and thus generates uneven the pressure drops over the plate. Consequently, the flow rates are reduced, especially during the filling of the last wells. Hence with time the resulting binding capacities increased (Fig. 18).

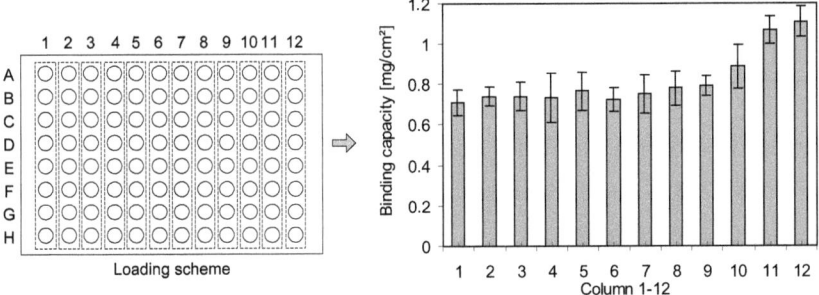

Figure 18: Effect of filling the wells without backpressure. 3 layers of IEX1 were used (1.02 cm² per well). Buffer for IEX1 was 20 mM Tris/HCl pH 7.4. After equilibration with 2.0 ml buffer per well, 2 ml BSA were added in two steps (amount 2.5 mg/cm²). At a permanent differential pressure of 350 mbar column 1-12 were loaded sequentially using the 8 pipetting tips (A-H). The overall process time was 240 s. The binding capacities were calculated by analysing the flow-through fractions.

4.2.3 Applications in high-throughput format

This chapter specifically deals with applications of chromatographic methods using the HTS system. The most prominent application is the load of a substance to screen for appropriate binding, washing or elution conditions. Among others, this small-scale approach can be used for chromatofocusing. This technique separates proteins due to their isoelectric point [92, 93]. For example, the protein is dissolved in buffers of different pH, e.g. stepwise 0.5 pH values. Using an anion or cation exchanger membrane the different protein containing solutions will be filtered through separate wells. By analysing the flow-through fractions the appearance indicates the region of the isoelectric point of the protein.

A further refinement would be the analysis of protein mixtures or real process solutions. By analyzing the breakthrough fractions the separation of several substances can be evaluated. Furthermore, in bind and elution mode wash and elution steps could be improved by screening for conditions of separation different molecules or groups of proteins. This application and the interpretation of results are considered more in detail in Chapter 5.

Another issue is the possibility to generate breakthrough curves and binding isotherms. This is feasible by selecting suitable load steps in 96-well format: an example is given in Figure 19.

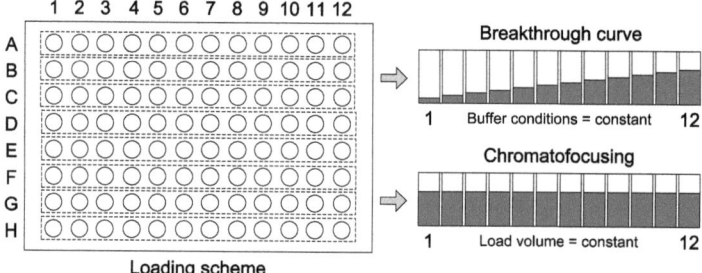

Figure 19: Example loading scheme for chromatofocusing and breakthrough curves. For 8 breakthrough curves the amount of loading volume would be increased, e.g., from 400 to 2000 µl. For chromatofocusing of 8 proteins (A – H) the loading pH would be changed from columns 1 – 12.

The breakthrough curve describes the time course of an adsorbent during the loading of a solution (feed) with a constant composition of molecules. The concentration gradient of the permeate is measured. In general, the breakthrough begins if the concentration of the molecule of interest in the permeate is > 0. Until then, the binding capacity corresponds to loaded amount. If the concentration c in the permeate corresponds to the initial concentration c_0, the saturation of the binding sites is achieved (saturation). In advance of the saturation capacity, the dynamic binding capacity can be determined. This value puts the binding capacity in regard to the fraction of the permeate to the initial (feed) concentration. In chromatography is the determination of the dynamic binding capacity at 10 % breakthrough a typical benchmark for the performance (Fig. 20).

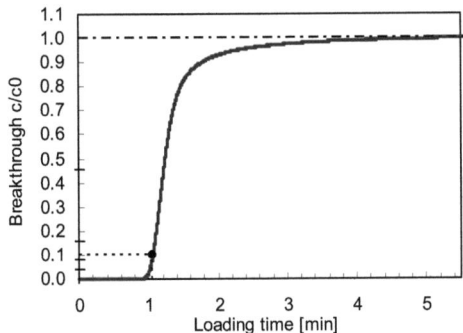

Figure 20: Generic breakthrough curve of a membrane adsorber. With increasing amount of loaded target molecule the saturation of the adsorber is achieved. The ratio of concentration in the permeate to feed tends to 1 (static binding capacity). The ratio of 0.1 identifies 10 % breakthrough (dynamic binding capacity).

The binding capacity behaviour BC (mg/cm²) of a membrane adsorber is described by

$$BC = \frac{\dot{V}}{A}\int_0^{t_{BT}}(c_0 - c_t)dt \qquad (4.11)$$

Where t_{BT} is the loading time of the target molecule with the initial concentration c_0 and c_t the concentration in the flow-through at a particular time. The binding capacity depending on the load density of the target molecule is calculated by

$$BC = \frac{1}{A}\int_0^{V_{Load}}(c_0 - c_{Flow-Through})dV \qquad (4.12)$$

Where V_{Load} is the loaded volume of the target molecule solution with the initial concentration c_0 and $c_{Flow-Through}$ the concentration of protein in the flow-through with the volume V_{Load}.

At the level of the saturation the adsorption equilibrium is reached. In chromatography, isotherms describe the distribution of components between the stationary (membrane adsorber) and mobile phase (buffer). Among others, the maximum binding capacity as a function of the component concentration can be determined depending on various buffer conditions and additives. The adsorption equilibrium can be described by using various models. At sufficiently low concentrations non-linear isotherms can be approximated by linear approaches following Henry, under assumption of infinite binding sites [129]. For chromatographic processes, however, the non-linear range is crucial as the most effective utilization of the adsorbent is to be achieved. The Langmuir isotherm is one model to define non-linear equilibrium reactions. Simplified assumptions are a reversible adsorption and no lateral interactions between different molecules and the formation of a monolayer [77, 86, 130].

$$q = \frac{q_{max} \times K \times c_0}{1 + K \times c_0} \tag{4.13}$$

The binding capacity is described as follows: c is the concentration of the component in the mobile phase, q_{max} the maximum binding capacity, K is the equilibrium constant and q the existing capacity at equilibrium. K is the thermodynamic equilibrium of adsorption and desorption rate. By dynamically adding a sufficient amount of protein solution the equilibrium be achieved for breakthrough curves, c/c_0 is then 1. This can be done with different concentrations and generates values for the calculation of the isotherms. In contrast to static methods like shaking the adsorbent in an e.g. protein solution, the dynamic filtration accelerates the process since the molecules reduces the binding sites faster [131].

4.2.4 Scalability of binding performance to larger devices

A general problem in chromatographic applications is the transfer of the performance between different scales. This mainly concerns small-scale approaches. The scalability of the 96-well plates to larger devices has to be evaluated. In regard to different chromatographic conditions data for 96-well plates and for lab scale devices will be presented in the following. The MA15 device consist of three layers membrane adsorber clamped into a polysulfone housing with a total membrane area of 14.9 cm^2 and a bed height of 0.8 mm. The void volume was determined with approximately 1 ml.

A parameter often used to benchmark chromatography matrices is the binding capacity, expressed as the maximum binding capacity or capacity at 10 % breakthrough. Buffer composition, flow rate and protein concentrations have influence on the adsorption kinetics and equilibrium. In general, these effects are unknown for the given system i.e. adsorber and target when starting a screening. The following example illustrates the issue. Different buffer compositions are applied to find binding conditions with high capacity for a certain contaminant. A fixed amount of protein (load) is filtered at these conditions. The binding capacities are calculated from the mass balance of the initial concentration and flow-through fraction in each well. From the calculated capacities one cannot deduce whether the adsorption equilibrium is reached. Figure 21 illustrates the situation for several breakthrough curves. Running only a single loading does not enable finding the maximum binding capacity at a given buffer condition.

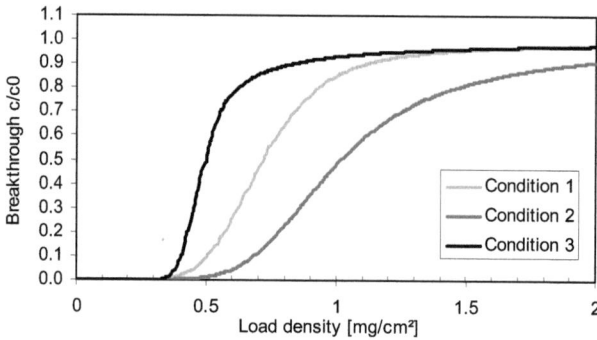

Figure 21: Schematic breakthrough curves at different buffer conditions. IEX2 was built in an MA15 device with 3 layers and 15 cm² membrane area. Three runs were performed at different buffer conditions. The saturation of the membrane is achieved at different amounts of protein applied (load density). At given values of load the breakthrough value c/c_0 differs. If the breakthrough is determined only at a fixed load no statement about the saturation of the membrane can be made.

Taking into account the issues discussed above and the fact that the residence time using 96-well plates is much lower than in larger devices, the scalability of maximum binding capacity has to be addressed critically. Commercially available anion exchange chromatography devices have been used to evaluate the scalability of binding capacity. The binding capacities of the two systems were compared under equal protein concentration, buffer conditions and amount of protein loaded per membrane volume. The calculation of the binding capacity using MA15 is described in the Appendix, for the 96-well plates a modified formula 4.3 was used (4.14). To increase the precision of the determination of the void volume V_V of the 96-well membrane holder was inspected more closely. The permeate side had a hold up volume of 46 µl. Because of a very low bubble point of the membrane below 0.3 bar this volume was always still filled with liquid after the equilibration and loading steps.

$$BC_{FT} = \frac{\left(c_{Load} - c_{Flow-Through} \times (1 + \frac{V_V}{V_{Load}})\right) \times (V_{Load} - V_V)}{A}$$ (4.14)

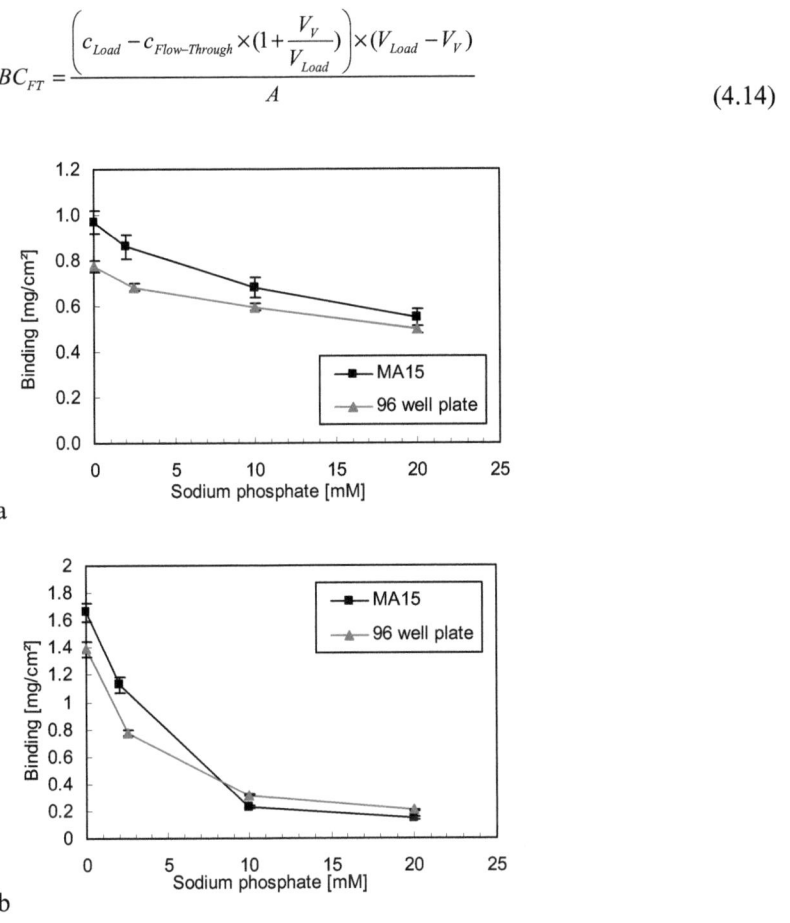

a

b

Figure 22: Comparison of protein binding. Both devices had 3 layers of IEX1 (a) or salt-tolerant IEX2 (b). Buffer was 20 mM Tris/HCl pH 7.4. Using the automated buffer mixing procedure, different concentrations of phosphate were added to the buffer and protein solution as indicated in the figures. After equilibration with 2.0 ml buffer per well, 2.0 mg/cm² of BSA were added in 2 steps of 0.5 ml each. For a direct comparison the same types of membranes assembled into a MA15 device with 3 layers and 15 cm² membrane. The loading volumes were normalized to the membrane area.

The difference between the two graphs could be explained by the influence of the factors described above. Furthermore, differences due to fluctuations in the production of different membrane lots could not be excluded. Nevertheless, a clear trend for both membrane types is evident. In Figure 23 the scalability of the relative values is presented.

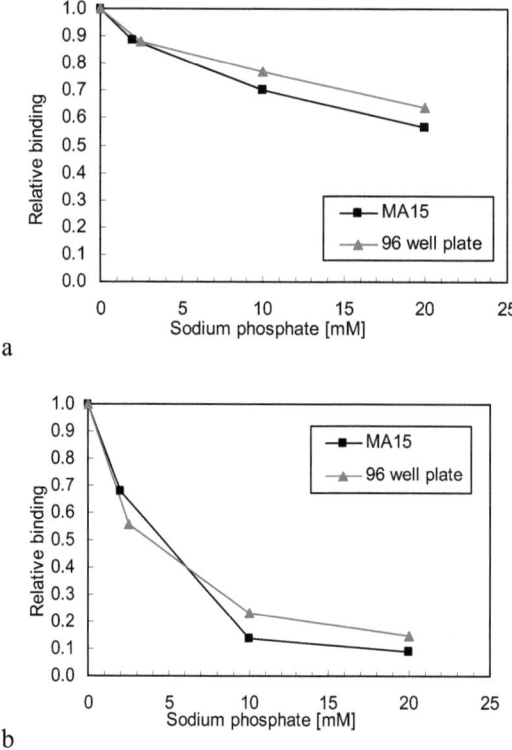

Figure 23: **Comparison of relative values in protein binding.** The data from Figure 22 are compared relative to the maximum value for each IEX (a: IEX1, b: IEX2).

The data are generated under conditions of overloading. That means less binding sites were available to bind all molecules. The breakthrough is > 0 and the loading reaches the adsorption maximum.

In further trials the binding properties of phage ΦX174 have been investigated. In this case the amount of free binding sites in the adsorbing matrix exceed the amount required for binding the phage particles several fold as confirmed in the Appendix (8.2.7). The results show comparable relative values at different buffer conditions for both devices (Fig 24 + 25).

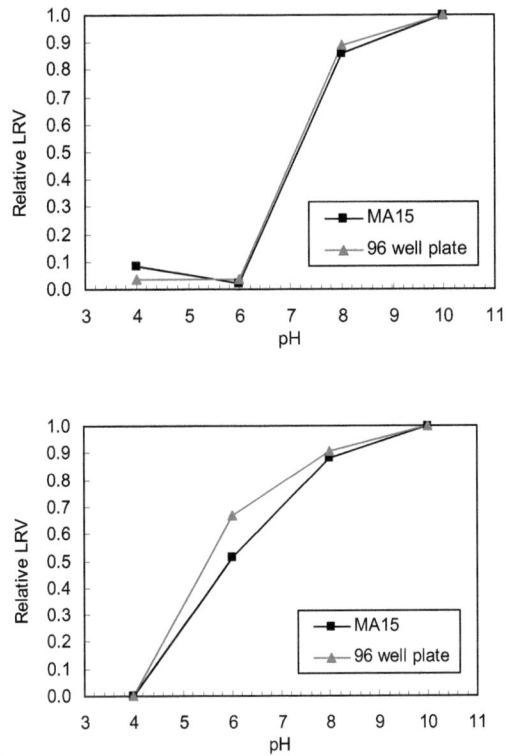

Figure 24: Comparison of phage binding. Both devices were equipped with 3 layers of IEX1 (a) or IEX2 (b). Buffers (20 mM) were at pH 4: NaAc, pH 6: BIS-Tris, pH 8: Tris/HCl and pH 10: CHES. Using the automated buffer mixing procedure, different concentrations of phosphate were in the buffers and protein solutions. After equilibration with 2.0 ml buffer per well, 2.0 ml of phage solutions were added. The initial infectivity titer was 1.5×10^7 PFU/ml. For comparison the same types of membranes were built-in a MA15 Device with 3 layers and 15 cm² membrane. The loading volumes were normalized to the membrane area. The absolute LRVs were compared with the maximum value for each membrane.

In the following the scalability of protein binding depending on the loaded amount of protein using the 96-well membrane holder in comparison with a larger device is shown. As previously discussed, the short contact time generates earlier breakthrough.

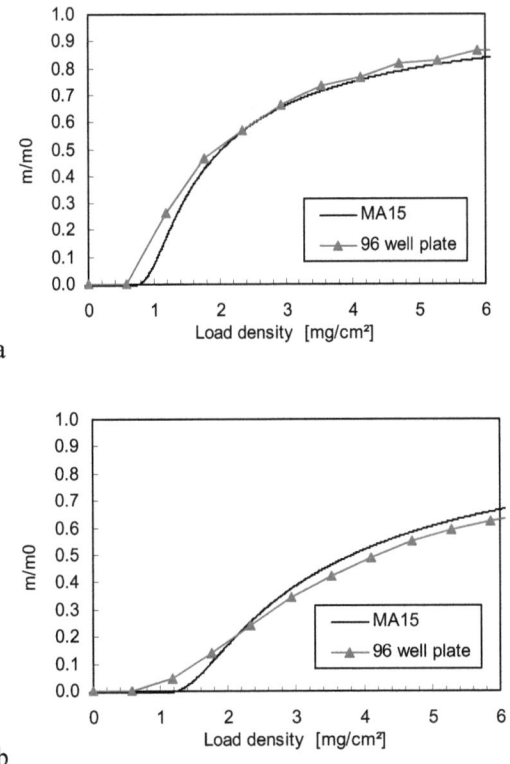

Figure 25: Comparison of the breakthrough behaviour. Both devices were applied with 3 layers of IEX1 (a) or IEX2 (b). Buffer was 20 mM Tris/HCl pH 7.4. The breakthrough behaviour is described as the ratio of the amount m of protein in the flow-through to the loaded amount m_0. The breakthrough curves with MA15 devices were determined using an ÄKTA FPLC system described in the Appendix. The flow rate was 15 ml/min (3 MV/min). For both experimental setups the initial BSA concentration was 1 g/l. The amount of protein in flow-through of the 96-well plates was analysed by measurement of the concentrations of BSA. The loading scheme was similar to Fig. 19. For a load volume > 2 ml the fractions were pooled before measurement of the concentration in the flow-through.

The data show the scalability of relative values from 96-well to the MA 15 devices. Due to the influencing factors and the errors due to small working volumes the transfer of absolute values from 96-well to larger scales is limited.

4.3 Conclusion

The first part of the thesis dealed with the problematic of needed improvements regarding time and material consumption and the investigation of complex process steps in chromatographic applications using membrane adsorbers.

The development of a 96-well membrane holder [125] simplified the incorporation of modified membranes and the evaluated liquid-handling system enables the parallel processing of a large number of small samples of buffers and protein solutions. The three membrane layers used in this study, respectively the membrane area per well lead to optimized material consumption, required sample size and improve process stability.

The screening applications presented here highlighted critical factors to be considered during operation. An optimized setting of liquid-handling parameters and work flow reduced the possible error. In spite of the potential heterogeneity of the pore structure and ligand distribution of the membrane, a negligible standard deviation in the well-to-well comparison was observed. A limited scalability from 96-well formats to larger devices has been shown. The loading amount is critical when scaling the membrane to larger devices, e.g. to determine the maximum binding capacity the saturation of all binding sites have to be assured by loading a sufficient amount of protein to a well. In particular, the high and limited controllable flow rate is of most importance. Different model systems have confirmed the assumption that higher flow rates cause a considerable increase the passage of protein molecules passing the membrane. This was especially true for very small amounts of protein in the breakthrough. However, it was confirmed that the relative ratios between different buffer conditions can be scaled up.

The establishment of a high-throughput screening system for membrane chromatography leads to new scopes of operation. Due to the high degree of parallelization it was possible to perform a high number of experiments compared to conventional fast protein liquid chromatography which would not allow such a high resolution of influencing factors on a broad range. The use of a statistical experimental design can reduce the experimental effort further. However, this will increase the risk of failing to spot interactions between involved factors by necessarily reducing the generated data.

After the establishment of the screening system, in the next chapter the investigation of anion exchanger membrane adsorbers for the separation of contaminants from the target molecule was performed.

5 Investigation of contaminant removal using anion exchanger membranes at various process conditions

Chapter 4 of this work focused on the development of an automated HTS system. This platform now has been used to characterize the contaminant removal by using positively charged membrane adsorbers. Due to the previously limited available small scale approaches, published studies are often limited regarding the investigation of the binding performance of membrane adsorbers using a large number of buffer conditions. The studies often considered only a small number of conditions in a small operational window. Furthermore, due to different properties of the used model or process solution like concentration and purity, various process parameters like flushing volumes or flow rates and analysis also a comparison of data from different publications is limited. Therefore another aim of this thesis was a systematic evaluation of the binding performance of anion membrane adsorbers under strictly comparable buffer conditions. Influencing factors on the binding and separation performance have been evaluated. Model systems consisting of single or multiple proteinogenic components were used. The selection of buffer conditions was based on preliminary results and published data. Furthermore, the thresholds of common process conditions for anion chromatography were considered.

Previous results have been shown that especially due to the dependence on the residence time the comparison in binding of 96-well to larger devices is somewhat limited. Therefore, the focus was on the interpretation of relative changes of binding performance due to the buffer conditions which were investigated.

5.1 Model systems for studying chromatography in downstream processing

The understanding of the mechanisms of contaminant removal by chromatography media during the purification in downstream processes requires reproducible model systems to mimic real process solutions. The use of real process media does not necessarily generate more meaningful data for analysis, due to a more complex composition of the medium and interactions of components which can adversely affect the interpretation of results. Using the same cell line for different products is preferred for the production process. The genetic information of the master cell line is only slightly modified to express a specific target molecule. For example, in mAb production based on the same cell line it has been shown that the purification performance of a chromatographic step varies for various antibodies [110, 133]. Small variations during repeated cultivations can change the composition of HCP or the amount of target protein. Factors like stability due to the storage of real process media or buffer exchange have to be controlled carefully. These issues can be reduced by using model systems. Also process development approaches with model systems are described. A common method is the use of spiking solutions, for example, containing viruses or endotoxins for contaminant removal studies [59, 84]. Depending on the purity of the spike solution, further contaminants can be introduced into the process media due to the incomplete purification of the virus spike.

Within this thesis the approach was to establish model systems. It was assumed that knowledge of these processes can later be transferred and compared to real processes. The chosen molecules, which have been used, are presented in the following.

The production and use of infectious viruses as a model is complicated, time consuming and associated with safety concerns regarding the health risk of the operators [93, 134]. Therefore, bacteriophages are an accepted alternative as a model system with several advantages: They can be produced in microbial organisms with short cell cycles like *Escherichia coli* C (*E.coli*) and are generally harmless to humans because they infect only bacteria. Disadvantages are that phages are smaller than most viruses and have different surface proteins, which can affect the binding to an adsorbent. Nevertheless, several studies have shown the comparability to viruses and support the use of bacteriophages as appropriate model viruses [84, 95]. For the investigations presented here phage ΦX174 was used. Other commonly used models are PP7 and PR772 [93]. The icosahedral shaped bacteriophage ΦX174 has a diameter of 25-30 nm [135, 136]. The isoelectric point pI is reported between 6.4 and 6.7 and has a thin protein coat covering a nucleic acid core [135, 137]. It does not have a lipid envelope. It consists of 26 % nucleic acid [135] and, furthermore, has a protein density of 1.4 mg/ml [93]. The phages were produced by infection of cultivated *E.coli* cells. Production and spike preparation is described in the Appendix.

Host cell proteins are perhaps the most inhomogeneous class of impurities. Different components were described in different cell cultures. Often used Chinese hamster ovary (CHO) cell lines were studied regarding the composition of HCP by Jin, M. et al. [138]. The distribution in terms of molecular weight and pI was shown. The largest proportion of HCP had a pI between 5 and 6, most proteins had a molecular weight between 25 and 75 kDa. Similar measurements were performed by Matthias Kraetzig during his Master thesis [139]. The analysis using two-dimensional gel electrophoresis identified proteins with a molecular weight of 25 to 60 kDa. Due to similar distribution of HCP in real process solutions regarding pI and size, several commercially available proteins can be used to mimic HCP solution. BSA, lysozyme, myoglobin, ovalbumin, conalbumin, cytochrome c and chymotrypsinogen were described as well-fitting model proteins [78]. In the work presented here BSA and Green Fluorescent Protein GFP were used as model systems. BSA has a pI of 4.7 and a molecular weight of 66.4 kDa [63, 100], it binds to anion exchangers at neutral pH. This also applies for GFP with a pI of 5.8 and 29 kDa. GFP was chosen to mimic a smaller HCP. Furthermore, the advantage of the protein was the simpler analysis by measuring its fluorescence. The calibration curve displayed in the Appendix confirms a possible detection even in minute quantities. Additionally to the use as contaminants, these proteins can also mimic a target molecule during the process development.

Likewise to HCP, the molecular weight of host cell DNA can have a broad range. The size of DNA and fragments are expressed as the number of its base pairs (bp). Salmon sperm DNA, with a MW of approximately 1000 bp, was used as a smaller; calf thymus DNA with a MW up to 24000 bp was used as a larger model contaminant. To evaluate the endotoxin removal, the Limulus Amebocyte Lysate test was used as described in 8.2.6. It is derived from the blood cells of Limulus polyphemus, the Atlantic horseshoe crab [140].

In general, these individual components can be used systematically to characterize adsorptive materials. This applies to the use individually, as well as a mixture of some or all components. From the generated data, information about the performance of adsorbers like binding capacity, recovery and reuse after cleaning can be drawn. From this, it may be possible to transfer the results to real process solutions. This necessitates the right choice of model systems.

5.2 Experimental design and evaluation of results

The high-throughput screening approach requires the planning and evaluation of appropriate tests. For the investigation of a large number of process conditions a statistical design of experiments can be used. The principle behind is to reduce the number of measurements and for experiments to determine influencing factors. The selection of the chosen experimental conditions and representation of results has been explained in this section.

The necessary numbers of measurements e depends mainly on the number of variables k like pH or conductivity in chromatography. The different setpoints n_v for each variable affect the effort.

$$e = n_v^k \qquad (5.1)$$

Often $n_v > 2$ is required to minimize to the loss of information or to use non-linear regression models. Thus, the number of trials grows faster than the number of variables. Therefore, using a reduced experimental design reduces the effort, like the central composite design or the Box-Behnken design. A disadvantage of the concept of experimental designs, investigating a large number of variables with only few measurements, is the risk of misinterpretation of the data or a low amount of evidence [108]. The evaluation of the experimental design can be determined via the power, depending on the signal-to-noise ratio. Noise is designated as the standard deviation. Signal is defined as the smallest change of a target value, like binding capacity, recovery or separation in chromatography. Effects can be determined using mathematical relationships to represent correlation. Significant factors are ascertained from variance analysis like one-way-ANOVA [141, 142].

The advantage of the high-throughput approach is the possibility to run a high number of experiments parallel. The experimental designs selected in this thesis were based on existing knowledge from previous experiments. These included the known conditions for chromatographic steps in downstream processing or the isoelectric point of proteins. Thus, the density of data points for conductivity in known condition areas was higher than in the border areas of possible applications. The pH was set around 7. This is one common setpoint in the purification of monoclonal antibodies by ion exchanger chromatography, e.g. to prevent aggregation [9, 64, 65].

Large data sets require the use of a presentation that allows visual interpretation with programs for statistical analysis. The dependence of a target value from two parameters can be presented by three-dimensional plots. For the sake of clarity, the three-dimensional plot is transformed into a two-dimensional contour plot. The target value is mapped using a color gradation. This form of presentation was preferred in this thesis. The color surfaces between the individual checkpoints were determined by a linear interpolation (for the algorithm used see the Appendix).

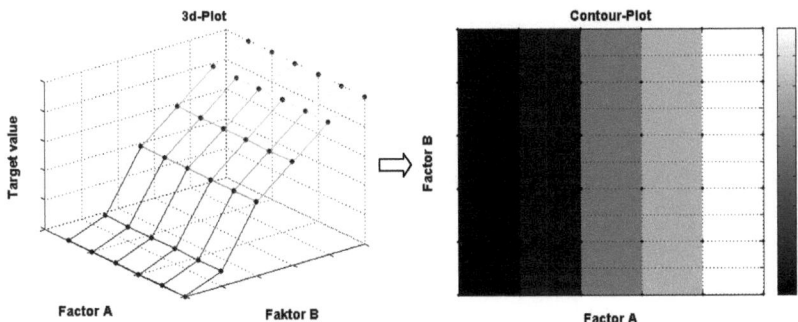

Figure 26: Example of a contour plot. The target value depends on influencing factors A and B. Conversion of the three-dimensional plot (left) into a contour plot (right). Black dots mark measured data. The target value is mapped using a color gradation. In this example the target value only depends on factor A.

Nevertheless, this kind of presentation is limited to three dimensions. Extension to four dimensions by a three-dimensional plot and combination with a color gradation lead to unclear arrangements of data. For the presentation of more influencing factors and target values, several contour plots can be combined. Sweet spot analysis considers the different variables in downstream processing. These are a superposition of several contour plots. In purification using chromatography yield and the amount of contaminants like aggregates, DNA and HCP are of interest. Figure 27 shows a sweet spot analysis. It displays the knowledge area, control space and design space of a process [96, 143, 144] in a single figure.

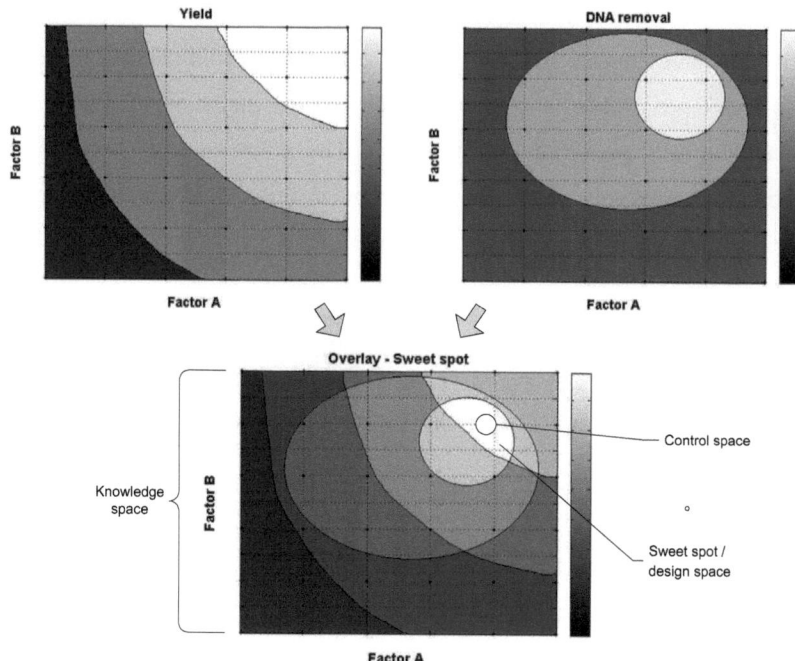

Figure 27: Example of a sweet spot analysis. Factors A and B affect both the yield of target and DNA removal of a purification step (upper part). By superposing of both charts the preferred working area is determined. Knowledge space includes the investigated process conditions. In the sweet spot, also known as the design space, the necessary conditions for that step are contained. The control space is located within this design space. The control space specifies the acceptable limits for the influencing parameters A and B during this process step. This method enables the inclusion of several variables into the analysis.

Sensitive methods for the analysis of trace concentrations of contaminants are used in DSP. Assays based on enzymatic reactions are particularly sensitive and easy to disturb e.g. the dependence of the endotoxin detection assay on pH and proteins [58]. The PicoGreen assay for the detection of DNA is distributed by components like BSA. To circumvent this, appropriate assays have to be used or complex multivariate calibrations are necessary [145]. In this thesis different approaches have been used:

- Model systems whit only a single component

- Mixed-model solutions with different assays, wherein the detection method for a component was negligibly affected by the others

A further approach was the analysis by comparing concentrations of the initial feed solution and the permeate. Based on the assumption of the identical effect of the buffer the breakthrough was calculated. An example was GFP, whose fluorescence depends on the pH value [146]. With a constant pH, the breakthrough of the component could be determined in a broad concentration range.

5.3 Binding performance of anion exchanger membrane adsorbers

Due to binding capacities and very high flow rate, membrane adsorbers are particularly suitable for final purification steps (polishing) where only small amounts of contaminants are present. Since typically anion exchangers are used, the knowledge regarding performance and interactions is important. Membrane adsorbers containing quaternary ammonium ligands have already been studied to some extent. Besides them, salt-tolerant anion exchangers are particular of interest. They bind biomolecules at higher salt concentrations or conductivities. Before the anion exchanger step, the conductivity of protein solution has often to be adjusted. A major advantage of salt tolerance is the reduction of buffer needed, because no or less dilution is necessary. A lower dilution i.e. smaller process volume also reduces the time for subsequent concentration after the chromatographic step in flow-through mode [33].

5.3.1 Influences of salt and pH on the binding of model contaminants

Generally, electrostatic interactions are responsible for binding of molecules to ion exchangers. The binding performance is strongly affected by the presence of salt, the pH-value and the isoelectric point of the present molecules. However, further mechanism like weak hydrophobic interactions or mutual influence of components can affect the performance of an anion exchanger [94, 147, 148]. So the first part of the evaluation of contaminant removal using membrane adsorbers the performance of anion exchangers at the binding of single model molecules were performed. Factors were pH, mono- and multivalent salts. In contrast to monovalent, multivalent salts consist of more than one electron to form an ionic bond with another molecule. It has been found that multivalent salts can induce effects on proteins like surface charge inversion [149] and affect the separation of proteins. It was expected that these effects could also be of interest for the purification in downstream processing using membrane adsorbers. Differences between the interactions of contaminants with both anion exchanger membranes should be examined. For all screening experiments the 96-well membrane holder consisting of 3 layers of membrane were used. The pressure was reduced to -350 mbar.

The first trace impurity studied was the phage $\Phi X174$. It was kindly donated by the R&D department at Sartorius Stedim Biotech GmbH. The preparation is described in the Appendix (8.1.6). The salt tolerance of IEX2 was particularly shown using this model contaminant. The detection limit was determined to be LRV 5. The results are shown as contour plot in Figure 28.

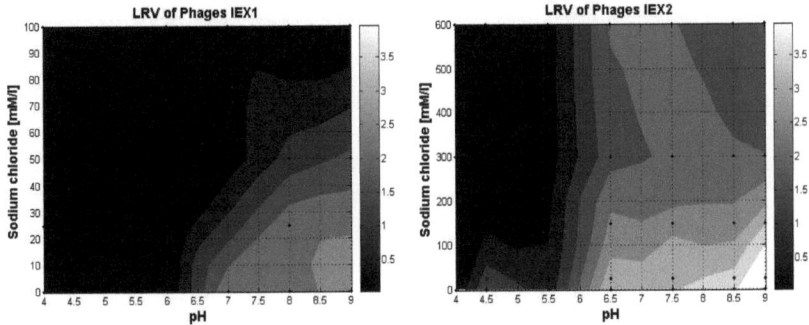

Figure 28: Influence of NaCl and pH on the binding of phages. Buffers for each IEX were 20 mM sodium acetate at pH ≤ 5.5, 20 mM BIS-Tris at pH 6.5 and 20 mM Tris at pH ≥ 7.5. Equilibration was done with 2.0 ml buffer per well. 3 parallel samples were pooled for each data point before phage detection using the infectivity assay. The initial infectivity titer was 1.5×10^7 PFU/ml. 1 ml phage solution was added at each condition. Gradation of the color scheme is LRV 0.5.

While the LRV of IEX1 was already close to zero at a sodium chloride concentration of 100 mM, IEX2 still retained phages up to a conductivity > 50 mS/cm i.e. > 500 mM. For both IEX the binding was decreased at lower pH. IEX2 still bounds phages near the isoelectric point. This may be an indication for a further separation mechanism like slight hydrophobic interaction or that there is still sufficient interaction between the ligand and remaining negatively charged surface proteins of the phages. It is described that primary amines take both electrostatic and hydrogen bond interactions [94]. This enhances interaction between ligand and target and is also rated as the reason for the salt tolerance.

The next contaminant investigated was endotoxin. The binding was checked in the range of 0 to 500 mM sodium chloride and a pH range from 6 to 9. The pH was limited because both salt and pH are inhibiting the enzymatic detection reaction. The binding performance was calculated according to equation 4.10. The detection limit was determined to be LRV 4.

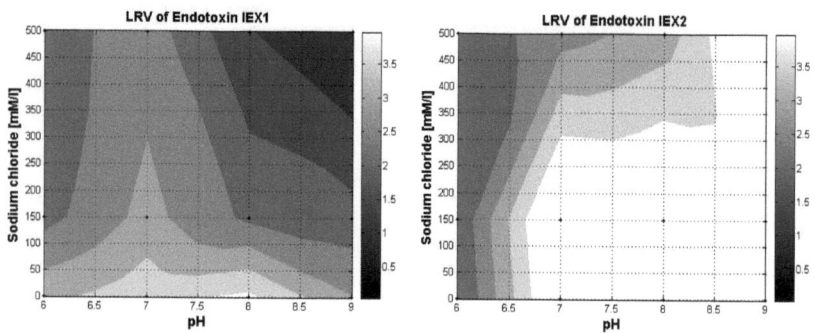

Figure 29: Influence of NaCl and pH on the binding of endotoxin. Buffers for each IEX were 20 mM BIS-Tris at pH ≤ 7 and 20 mM Tris at pH ≥ 8. Before equilibration each well was sanitized with 2 ml 1 M NaOH. Equilibration was done with 2.0 ml buffer per well. 12 different conditions were tested: three parallel samples each. 0.5 ml was added with a concentration of 500 EU/ml at each data point. Gradation of the color scheme is LRV 0.5.

Before using the 96-well membrane holder it was depyrogenized at 200 °C and the membrane was sanitized with sodium hydroxide. The results are shown as contour plots. Both membranes were able to remove the endotoxin over a broad operational range. The endotoxin is a lipopolysaccharide which consists of a lipid moiety and a polysaccharide that can contain phosphate groups. Responsible for the binding of endotoxin are the interaction between the negative charged phosphate groups of the protein structure and the positive charged ligands of anion exchanger media. Furthermore, it was mentioned that there are probably hydrophobic binding between the adsorbent and lipophilic groups of endotoxin too [59]. The log reduction level LRV was higher for IEX2. This membrane bound > 99 % of the endotoxins, even at higher salt concentrations. The lowest LRVs were found at pH 6 (LRV 2). The salt tolerance increases for higher values up to pH 9. IEX1 was more sensitive to higher salt concentrations, between pH 7 – 8 > 50 mM which correlates to published data [84]. The screening showed favored buffer conditions for endotoxin removal at higher salt concentrations around pH 7, where the endotoxin removal was still > 99 %. At higher pH-values and salt concentrations the LRV was below 2.

The binding of small salmon sperm and large calf thymus DNA to IEX1 and IEX2 has been examined. Sodium chloride was used to generate higher conductivities. Depending on the pH, appropriate buffer systems were used to ensure a sufficient buffering. The load concentration was 100 ng/ml. The concentration of DNA was measured by the PicoGreen Assay. The performances of binding the contaminant at different conditions were evaluated by calculating the removed amount of DNA. The 96-well plate was fully exploited: 24 conditions were tested, each using four parallel samples.

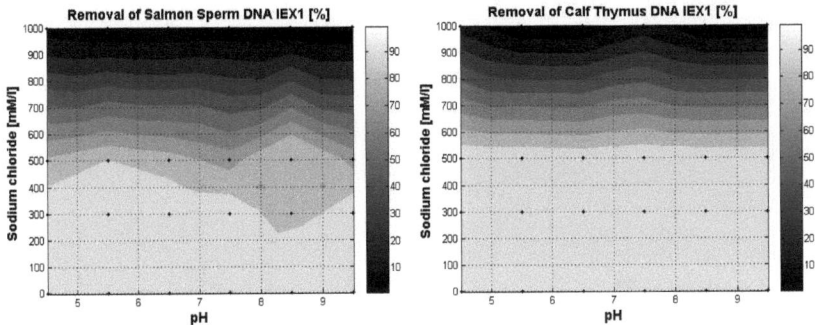

Figure 30: Influence of NaCl and pH on DNA binding. Buffers for each IEX were 20 mM sodium acetate at pH ≤ 5.5, 20 mM BIS-Tris at pH 6.5, 20 mM Tris at pH 7.5 – 8.5 and CHES at pH 9.5. Equilibration was done with 2.0 ml buffer per well. For each tested data point concentration was 100 ng/ml. 1 ml DNA solution was added at each condition. Gradation of the color scheme is 10 %.

The high affinity of DNA due to the negative charge caused a consistently good binding to the membrane. There was no significant difference in the removal of DNA at the different pH values tested. IEX1 showed a slightly higher breakthrough of the smaller Salmon sperm DNA. For all conditions the DNA concentration in the flow-through fraction was below the range of the calibration, where the lowest value was 10 ng/ml. The salt-tolerant IEX2 also showed no breakthrough at higher conductivities.

The presented results confirm trends due to buffer conditions, which were also described in the literature for other ion exchanger media [17, 84]. The amount of contaminants loaded was always much smaller than the binding capacity for that kind of molecule. Therefore the binding of BSA was investigated depending on the pH. The loaded amount of protein exceeded the maximum binding capacity of the installed membrane area.

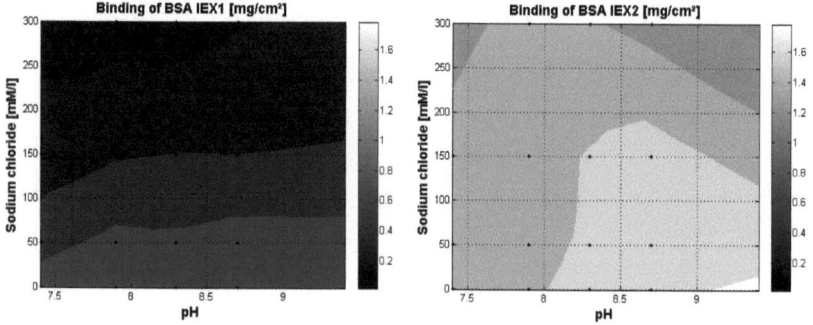

Figure 31: Influence of NaCl and pH on BSA binding. Buffer for each IEX was 20 mM Tris at pH. The pH was adjusted using HCl. Equilibration was done with 2.0 ml buffer per well. For each tested data point 4 parallel samples were analyzed. The amount loaded of BSA was 2.8 mg/cm². 1 ml BSA solution was added at each condition. The concentration of BSA was measured at 280 nm. Gradation of the color scheme is 0.2 mg/cm².

As expected, the binding capacity of IEX2 was higher than that of IEX1. The capacity of IEX1 slightly increased with higher pH. BSA was less bound for IEX2 at higher pH in combination with an increasing salt concentration. It was assumed that the loss of total binding capacity is caused by the reduction of protonated polyallylamine. Holmes-Farley et al. found that the protonation of polyallylamine will be reduced at pH > 8. Furthermore, they mentioned that at pH < 7 the ligand preferably appears as monoanion which also may reduce the binding forces [150].

In the following, the effect of multivalent ions on the binding behavior of BSA, GFP and DNA was investigated. Various amounts of phosphate or citrate buffer with the same pH were added to the basic buffer. Figure 32 shows the effect of sodium phosphate on the binding of Salmon sperm DNA to both ion exchanger matrices. Results are shown as contour plots.

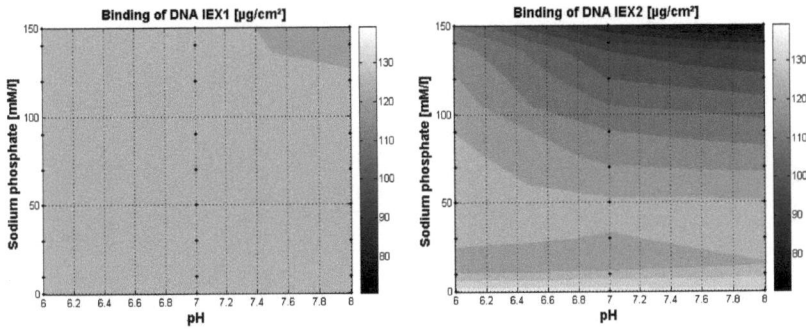

Figure 32: Influence of phosphate and pH on DNA binding. Buffers for each IEX were 20 mM BIS-Tris at pH 6 – 7 and 20 mM Tris at pH 8. Equilibration was done with 2.0 ml buffer per well. 3 parallel samples were analyzed for each tested conditions. The load density of DNA was 140 µg/cm². 1 ml DNA solution was added at each condition. The concentration of DNA was measured at 260 nm. Gradation of the color scheme is 5 µg/cm².

IEX1 was less affected by multivalent salt than IEX2. This showed that IEX2 is less tolerant towards multivalent ions. Nevertheless, also at higher concentrations of sodium phosphate binding of DNA occurred. As seen with BSA, the binding was reduced at higher pH values.

A noticeably binding of DNA at higher phosphate concentrations lend the interest to protein binding under such conditions. The binding of BSA was investigated over a broad range of pH and with different multivalent salts to detect differences to the removal of DNA. Both runs with phosphate and citrate were performed simultaneously testing 32 conditions at once. The results are shown as contour plot.

Investigation of contaminant removal using anion exchanger membranes

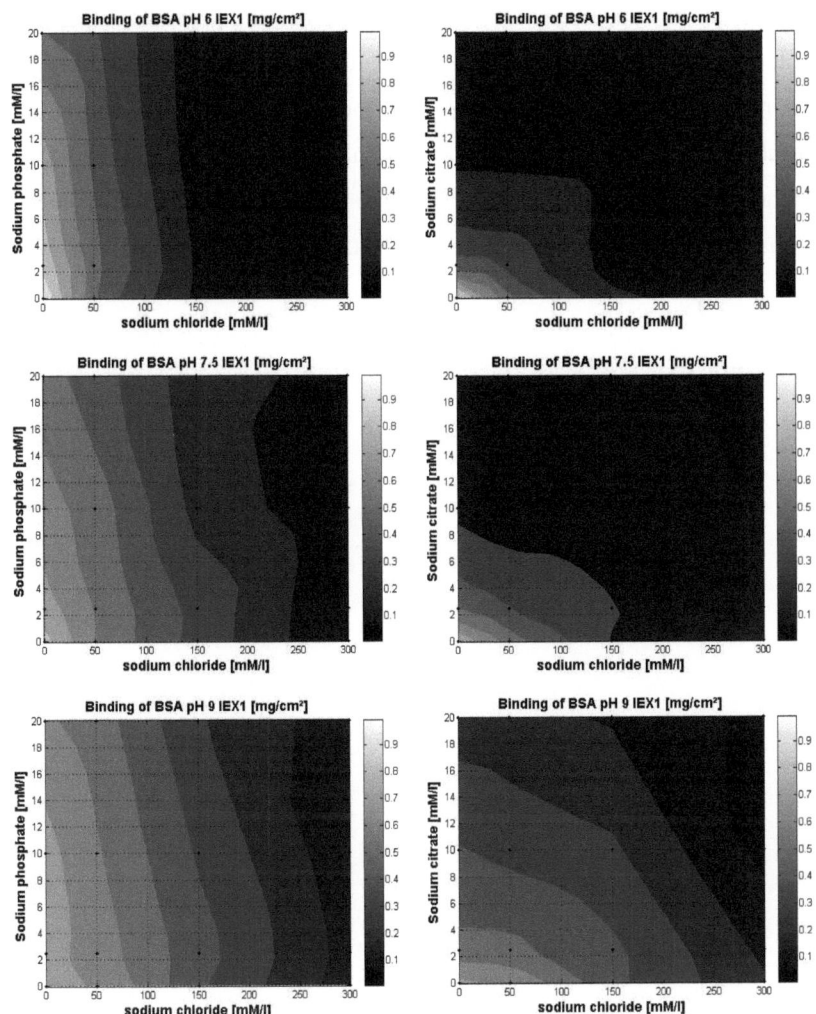

Figure 33: Influence of phosphate, citrate and pH on binding BSA IEX1.
Buffers were 20 mM BIS-Tris at pH 6, 20 mM Tris at pH 7.5 and 20 mM CHES at pH 9. Equilibration was done with 2.0 ml buffer per well. 3 parallel samples were analyzed for each tested conditions. The load density of BSA was 2.6 mg/cm². 1 ml BSA solution was added for each tested conditions. The concentration of BSA was measured at 280 nm. Gradation of the color scheme is 0.1 mg/cm².

Both mono- and multivalent salts reduced the binding performance of IEX1, while multivalent ions had a greater effect. Citrate decreased the binding more than phosphate. Furthermore, a higher pH increases binding with respect to the operational window. In Figure 34 the results are summarized for IEX2. In contrast to IEX1, multivalent salts decrease the binding while the monovalent sodium chloride showed no significant effect. Furthermore, at a pH of 6 and 9 a slightly increased conductivity affected the binding of IEX2 positively.

Investigation of contaminant removal using anion exchanger membranes

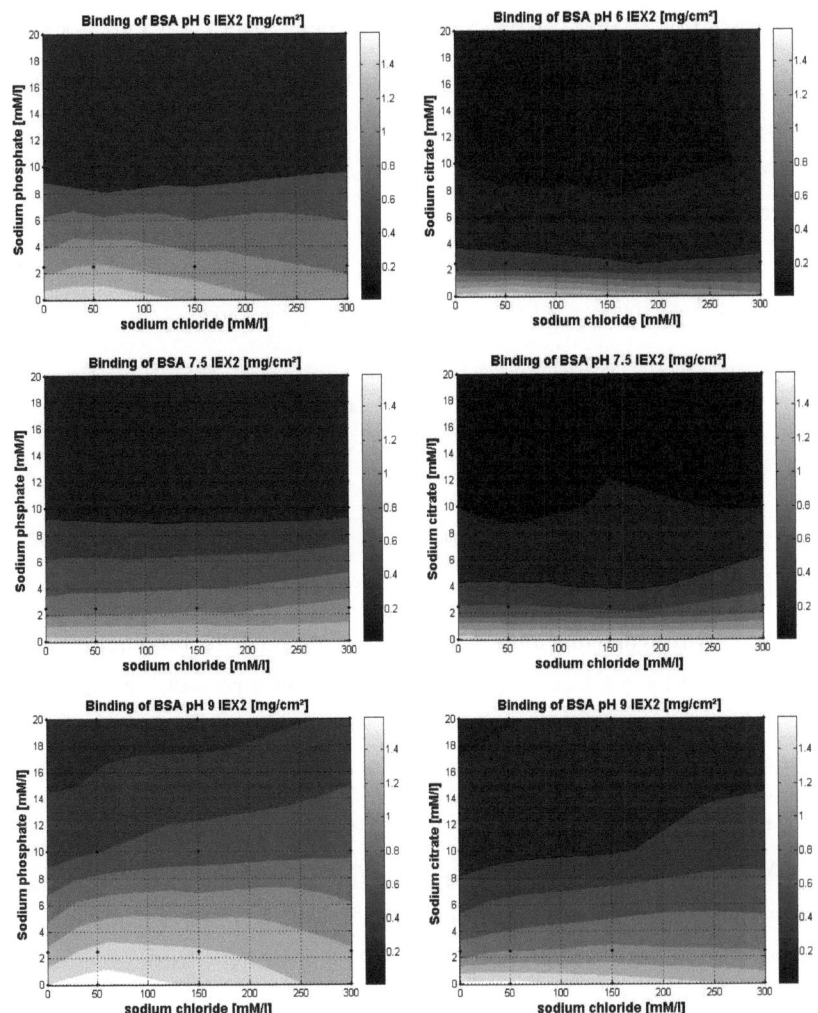

Figure 34: Influence of phosphate, citrate and pH on binding BSA IEX2.
Buffers were 20 mM BIS-Tris at pH 6, 20 mM Tris at pH 7.5 and 20 mM CHES at pH 9. Equilibration was done with 2.0 ml buffer per well. 3 parallel samples were analysed for each tested conditions. The load density of BSA was 2.6 mg/cm². 1 ml BSA solution was added at each condition. The concentration of BSA was measured at 280 nm. Gradation of the color scheme is 0.2 mg/cm².

A similar effect of phosphate was observed for low-concentrated protein with a pI in the same range like BSA. Green fluorescent protein at a low level of 1.6 µg/ml was loaded to the membranes at different conditions. The concentration was determined with the BCA reagent and expressed as BSA equivalents.

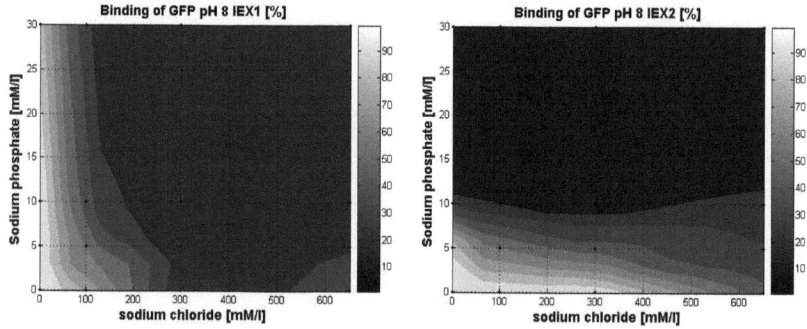

Figure 35: Influence of phosphate and chloride on GFP binding. Buffer was 20 mM Tris at pH 8. Equilibration was done with 2.0 ml buffer per well. 3 parallel samples were analysed for each condition. 1 ml GFP solution processed at each condition. Gradation of the color scheme is 10 %.

Both membranes retained the protein at low salt concentrations. Similar to the observations with BSA at a high load density, the effect of phosphate was much higher than chloride for IEX2. Thus, the binding behavior for phosphate > 10 mM and chloride > 200 mM differed and confirmed a significant difference between both anion exchangers.

Due to the large differences in binding capacity of IEX2 caused by the addition of multivalent salt ion, it is as apparently disadvantage of that type. In general citrate or potassium phosphate buffers reduce the purification power of these anion exchange membranes. However, it was assumed that this effect can be used for the separation of molecules.

5.3.2 Effect of mono- and multivalent salt on the separation of molecules

The previous result made the approach feasible to separate molecules by using multivalent salt ions without a strong increase of the conductivity. This could be an advantage in a purification scheme. The possibility was investigated using several model systems, which consist of at least two components. The performance of the membrane for separation a protein mixture was evaluated by the selectivity S. It was calculated by the ratio of the breakthrough of target molecule BT_{tm} and contaminant BT_{cont}.

$$S = \frac{BT_{tm}}{BT_{cont}} \qquad (5.2)$$

The minimum breakthrough value was defined as the lower detection limit. A high value for selectivity identifies preferred binding of the contaminant while the target molecule passes the membrane. If the selectivity $\ll 1$, the target molecule will be preferentially concentrated at the adsorbing media. Thus, the bind and elute mode is suitable for purification process. Process parameters, which ensure a high selectivity, however do not represent the most efficient operating conditions. Besides to selectivity the yield has to be as high as possible. If the removal of a contaminant is e.g. almost 100 % but also 20 % of the target molecule remains on the adsorbent, yield is not sufficient. The yields in the following experiments were given by the breakthrough of the target molecule calculated from the equal volumes of the load and permeate. Assuming that the capacity limit for a certain contaminant was not reached, a higher load can further improve the yield. Furthermore, studies have shown a displacement of target molecules by contaminants during a loading step [110]. The higher affinity of the contaminant for the binding sites displaced already bound molecules.

As a purification model process the removal of DNA from a mixture with GFP in flow-through mode was evaluated. The aim was to determine the conditions where the target molecule GFP passes the membrane whereas the DNA is retained. Due to their pI both components are negatively charged and interact therefore with positively charged anion exchanger. The DNA was measured photometrically at 260 nm; GFP by its fluorescence at 509 nm. The concentration of GFP was so low that the analysis of the DNA was not significantly affected. Conversely, the DNA showed no influence on the fluorescence. The separation of both components was examined in the range of pH 6 to 9, a sodium chloride concentration of 0 to 650 mM and a phosphate concentration up to 30 mM. The fluorescence of GFP depends on the pH [146] and was investigated in the Appendix. The results for pH 8 are shown in Figure 36. The minimum considered breakthrough value was 5 %; the maximum 95 %. Thus, the maximum predictable selectivity was 19 according to formula 5.2.

Investigation of contaminant removal using anion exchanger membranes

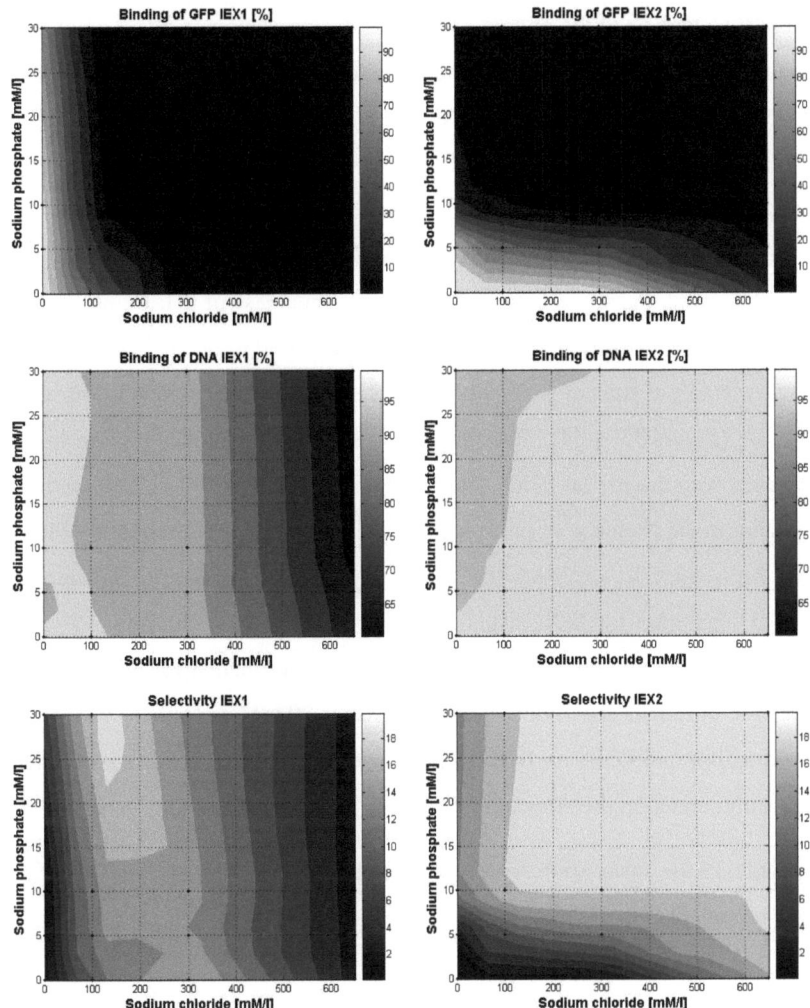

Figure 36: Separation of DNA and GFP. Buffer was 20 mM Tris at pH 8 [119]. Equilibration was done with 2.0 ml buffer per well. 3 parallel samples were analysed for each tested condition. 1 ml of a mixture of GFP and DNA was added at each condition. Gradations of the color scheme are 10 % for GFP, 5 % for DNA and 2 for the dimensionless selectivity (see formula 5.2). After the determination of the DNA and GFP in the flow-through fraction the selectivity was calculated.

The results confirmed the previous results with single components. The presence of multivalent salt for IEX1 had no effect; in contrast the binding of GFP to the adsorber was reduced on IEX2. Moreover, the binding of DNA was less affected by each salt composition for IEX2; it decreased for IEX1 at higher conductivities. The interaction of DNA with IEX1 was abolished with concentrations of sodium chloride > 100 mM. The addition of phosphate up to 30 mM showed no significant effect, similar the increase of pH. Using IEX2 the breakthrough of DNA increased slightly with the addition of phosphate. The complementary results at several pH values showed similar trends to those seen for BSA (Figure 33 + 34). The range of conditions for higher protein binding increased with the pH on IEX1. At pH 6 the affinity of GFP to IEX2 seemed to be higher in the presence of phosphate as well as chloride. Subsequently the data have been used to evaluate the selectivity. The visualization indicates the favored separation for IEX1 using 250 to 300 mM sodium chloride or lower in the combination with phosphate. The use of only phosphate up to 30 mM did not lead to a sufficient separation. Further, already 20 mM phosphate was able to separate GFP and DNA on IEX2. Complementary values for further pH values were summarized in the Appendix (8.3.2).

In Figure 35 the sweet spot, with requirements of 90 % GFP recovery and 90 % DNA removal is visualized.

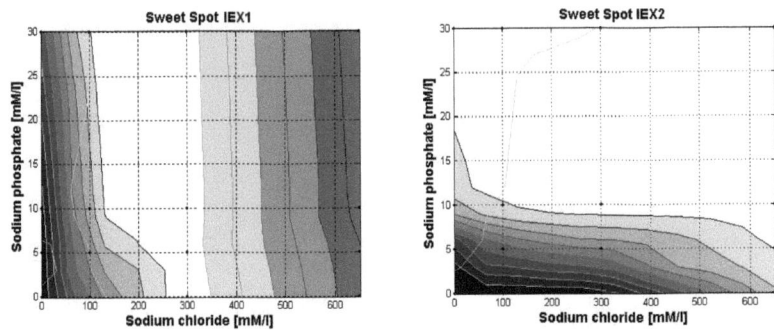

Figure 37: Sweet spot analysis separation DNA and GFP. Evaluation of the data from Figure 34. The brighter area highlighting buffer conditions where a sufficient separation is possible.

The high affinity of DNA to anion exchangers at the investigated conditions favors a purification step in flow-through mode. For both types of IEX appropriate conditions were found. A separation for IEX2 was possible with a slight increase of multivalent ions. Thus, it was not necessary to strongly raise the conductivity.

The results of the study using single model molecules showed less influence of the salt on binding of proteins and phages. Therefore, the following investigations focused on the purification performance of membrane to separate these molecules. Model solutions containing protein and phages were evaluated at different buffer conditions. The separation of GFP and phages was chosen to investigate the binding behaviour at low concentration of a contaminant and target molecule. In contrast to the binding capacity for BSA, the load used here was much smaller than capacity of the adsorber.

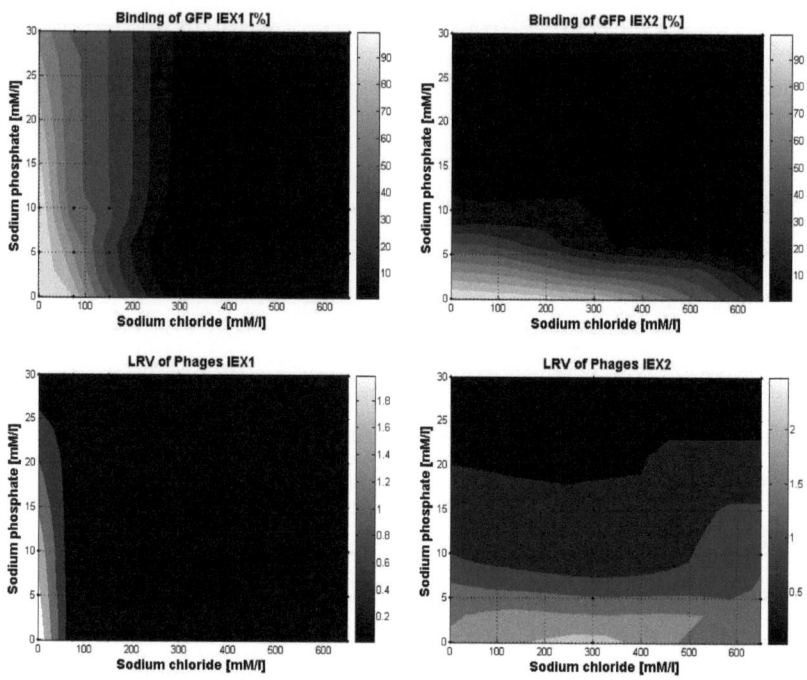

Figure 38: Separation of phages and GFP. Buffer was 20 mM Tris at pH 8. Equilibration was done with 2.0 ml buffer per well. 3 parallel samples were pooled and analysed for each tested condition. 1 ml mixed solution was added at each condition. Gradations of the color scheme are 10 % for GFP and 0.5 logs for phages.

Based on the results for IEX1 it was expected that no condition in the observed operational window would be useful to separate phages and GFP. But quite favorable process conditions could be found; wherein at least 95 % of GFP was found in the flow-through and > 99 % of phages were bound to the membrane. In contrast to the purification of GFP in flow-through mode, those conditions are also suitable in bind and elute mode. If GFP represents a typical HCP it passes through the membrane adsorber while the phages were bound to it.

DNA, GFP and phages was evaluated as suitable model components to evaluate anion exchangers in flow-through mode for contaminant removal. The detection methods for the biomolecules were independent of the components of this model solution. Phages, GFP and DNA as model components represent molecules of real process media regarding size and binding properties.

5.4 Comparison of the separation of molecules to larger device

With the use of the screening device process conditions have been identified which enable the separation of contaminants and target molecules. In Chapter 2.4.2 the scalability for IEX1 and IEX2 regarding the influence of pH to bind bacteriophage ΦX174 has been shown. Further data on the influence of sodium chloride using a larger device also confirm the possibility of scale up (see Appendix 8.3). In this part now the scalability of the separation of DNA and phages using multivalent salt was investigated.

A screening was performed using the 96-well membrane holder to evaluate the separation of a phage solution spiked with Salmon sperm DNA. The DNA concentration was 200 ng/ml.

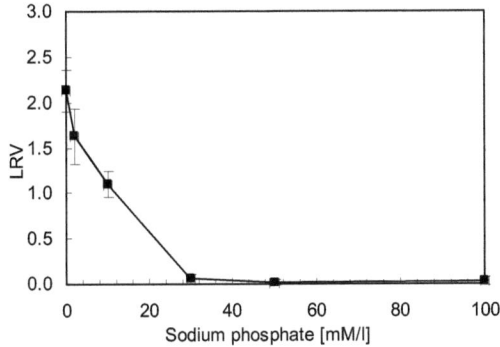

Figure 39: Separation of DNA and phages using phosphate. After equilibration with 2.0 ml 20 mM Tris buffer at a pH of 7.4, 1 ml of phage solutions was added to each well. Each buffer condition was testes 6 times. The initial infectivity titer was 1.5×10^7 PFU/ml.

30 mM sodium phosphate abolishes the binding of phage to IEX2 almost completely. The DNA content of each fraction was below the detection limit of 10 ng/ml.

The same device containing of 15 cm² membrane area was used according to the scale-up trials of Chapter 4.2.4. After equilibration using 50 ml, 150 ml of DNA-spiked phage solution was filtered at 10 ml/min. The buffer condition was similar to the screening: 20 mM Tris at pH 8. 30 mM sodium phosphate was added to separate DNA and phages. Three flow-through fractions of 50 ml were collected. For each fraction the DNA found to be below the detection limit of the test. The recovery of phages was 79 % analysed with the phage assay. The results confirmed the scalability of optimized buffer conditions for the separation of DNA and phages.

5.5 Conclusion

After implementation and testing the developed screening approach, the investigations in the second part of this thesis the platform was used for the characterization of contaminant binding using membrane adsorbers. Beside classical single molecule models like BSA or DNA, multi molecule model solutions were used to observe the contaminant binding properties of ion exchange membrane adsorbers regarding the effect of salts and pH on the performance. Furthermore, the binding of endotoxin and phages as trace impurities were investigated in a broad operational window.

Anion exchanger having quaternary ammonium IEX1 or polyallylamine IEX2 as ligand were compared. Depending on the analysed molecule, effects in the slightly basic and acidic pH range and up to 150 mM multivalent salts and 1000 mM monovalent salt were shown. Optimized buffer conditions for binding different molecules could be determined. Similar to published data, it was found that DNA and endotoxins have a high affinity to the anion exchanger media in a broad operational window for both membranes. The binding of phages, GFP and BSA was less affected at higher concentrations of monovalent salt using the salt-tolerant IEX2. The binding on IEX1 was strongly influenced by increased salt concentration even at a conductivity of 5 mS/cm. Multivalent salts have a greater effect on the binding performance for both membranes, but significantly stronger on IEX2. A lowering of protein and phage binding by the addition of multivalent salts compared to the addition of monovalent sodium chloride has been observed The effect of citrate was even greater compared to phosphate.

As expected, with IEX1 a higher pH, farer away from the isoelectric point, reduced the effect of salt. In contrast, this was not valid for the binding using the polyallylamine ligand of the weak anion exchanger IEX2. It was found that the binding in the presence of salt at slightly acidic conditions was improved. It was also possible to bind molecules even at or below the pI of the molecules using IEX2 as shown for phages and GFP.

Green fluorescent protein and phage Phi X 174 turned out to be a practical model system for investigation contaminant removal with low-concentrated target molecules. Well suited buffer conditions for the separation of DNA and GFP were successfully determined. The model system identified different approaches for contaminant removal using both anion exchanger membrane adsorbers. The separation using IEX1 was in general ruled by the conductivity. Using IEX2 the type of salt was crucial for the separation of contaminant and target molecules. The use of multivalent ions, opened approaches for the use of anion exchangers in flow-through as well as in capturing mode, although contaminants may have the same charge like the membrane surface.

Finally, it was possible to scale up and confirm the binding properties with larger devices. As well as for phage removal in Chapter 4, binding properties were successfully transferred using buffers which containing multivalent salts.

6 Summary and outlook

In the course of the work presented here a new screening platform has been developed to characterize the contaminant removal in downstream processing using membrane adsorbers. Thus, the investigation of the influence of up to 32 buffer conditions on chromatographic performance of membrane adsorbers was performed in 3 hours, including buffer preparation and analysis. The developed 96-well membrane holder for screening applications showed sufficient reproducibility and enables the simultaneous investigation of membranes at different buffer conditions without the need for additional sealing like O-rings or preparation of discrete membrane pieces [132]. Furthermore, the results and knowledge gained here have been used for the development and commercialization of a disposable three-layer 96-well screening device for screening approaches [151].

Investigations on the performance and the scalability showed that the results can be transferred to a larger scale. Figure 40 represents the favored route for generating robust, scalable and economic membrane chromatography purification steps.

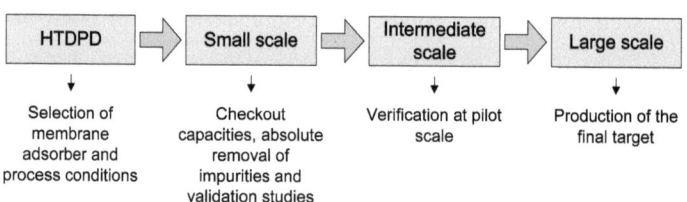

Figure 40: Route of process development using membrane adsorbers. Chromatography high-throughput downstream process development HTPD together with design of experiments will be used as a tool to determine the design and control space of purification steps.

The results of this thesis highlighted the need for screening tools for technology advancement in the field of membrane chromatography. It enables a large

Summary and Outlook

amount of process data to be generated and analyzed. This technique is of course not limited to ion exchanger membranes. For example, an obvious step is the investigation of hydrophobic membrane adsorbers for aggregate removal [13, 44, 115, 152, 153]. In respect to mixed-mode adsorbents with simultaneous different interaction modes of the ligand, it is of upmost importance to understand binding mechanisms behind. The increase in knowledge in membrane chromatography purification steps is probably the most important advancement. Chromatography is a major purification step in the production of biopharmaceuticals, therefore this screening approach can support the "Quality by Design" agenda of pharmaceutical regulatory authorities [138, 143, 144], wherein an aim is to facilitation of release and change criteria of process parameters. Together with new concepts of parallelized detection methods, like size determination of biomolecules [154], this further accelerates the accumulation of process knowledge. Further, screening results can be used for modeling chromatography steps and predicting purification performance. An improvement, due to availability of large data sets, is the possibility to use data-driven models. This kind of "Black Box" modeling [155] enables the prediction, even when physical and chemical relationships are rather obscure unknown, which is true for complex process media at different buffer conditions in chromatography steps.

This thesis reveals appropriate purification strategies in downstream processing. The present results showed an application which enables the concentration and purification of biomolecules. The removal of DNA fragments or host cell proteins in the purification of viruses or virus-like particles is of certain concern in purification [25]. An obvious application is the removal using anion exchangers at higher salt concentrations to remove DNA in flow-through mode. The use of multivalent salt in combination with polyallylamine based membrane adsorber has further the advantage of the possible purification in flow-through mode at low conductivity. Further studies could evaluate the concentration and purification of viruses or virus-like particles using anion exchanger. Due to the effect of small amounts of multivalent salts, it is expected that viral target molecules will bind while HCP passes through the membrane. The increase of the concentration of multivalent ions could then be used to elute the virus particles. In the same step, other contaminants with a high affinity to the ligand, like DNA or endotoxins, remain on the adsorbent. A patent application was applied to purify viruses with quaternary ammonium IEX1 and polyallylamine based IEX2 membrane adsorbers using the effect of multivalent ions [156]. Next to dead-end filtration, membrane adsorbers in tangential filtration mode are also described for the use in purification of biomolecules from particle containing media [85]. It could be evaluated to transfer the knowledge of buffer effects based on screening results also to this application.

Another interesting approach is the use of the salt-tolerant IEX2 to combine protein purification and desalting in a single step. Results in the thesis showed a binding of protein and phage molecules to the adsorber even at high monovalent salt concentrations. Prompting that multivalent ions can be used for elution using low-concentrated buffers the total salt concentration will be reduced. This might reduce further necessary steps in downstream purification. It combines a contaminant removal with a desalting step using an anion exchanger in bind and elute mode.

In the field of the development of newly modified membrane adsorbers the screening approach enables the implementation of methods for the rapid and parallel characterization of purification performances at several buffer conditions. The use of multi-molecule model solutions can be used to investigate parameters like the type of ligand, ligand density or pore size of the membranes. Thus, the development should not only be focused on binding capacity and the tolerances to buffer additives, but also on the effects on the selectivity.

7 References

[1] Varadaraju H., Schneiderman S., Zhang L., Fong H., Menkhaus T. J.: Process and economic evaluation for monoclonal antibody purification using a membrane-only process. Biotechnol Prog 2011 27(5), 1297-1305.

[2] Curling J., Gottschalk U.: Process chromatography: Five Decades of Innovation. BioPharm Int 2007 20(10), 10-19.

[3] Aldington S., Bonnerjea J.: Scale-up of monoclonal antibody purification processes. J Chromatogr B 2007 848(1), 64-78.

[4] Eriksson K., Ljunglöf A., Rodrigo G., Brekkan E.: MAb Contaminant Removal with a Multimodal Anion Exchanger. BioProcess Int 2009 7(2), 52–56.

[5] Berthold, W.: From High Dilution to High Titers - 35 Years of Antibody Manufacturing. Oral presentation at the 6th European Downstream Technology Forum, Goettingen (Germany), September 7-8, 2010.

[6] Morenweiser R.: Downstream processing of viral vectors and vaccines. Gene Therapy 2005 12, 103-110.

[7] Sommerfeld S., Strube J.: Challenges in biotechnology production - generic processes and process optimization for monoclonal antibodies. Chem Eng Proc 2005 44(10), 1123–1137.

[8] Sinclair A.: How to Evaluate the Cost Impact of Using Disposables in Biomanufacturing BioPharm Int 2008 21(6), 26-29.

[9] Collier R.: Rapidly rising clinical trial costs worry researchers. CMAJ 2009 180(3), 277-278.

[10] Vázquez-Rey M., Lang D. A.: Aggregates in monoclonal antibody manufacturing processes. Biotechnol Bioeng 2011 108(7), 1494–1508.

[11] Laursen I., Teisner B.: Process for Producing immunoglobulin for intravenous administration and other immunoglobulin Products. Patent US 6281336, 2001.

[12] Follman D. K., Fahrner R. L.: Factorial Screening of antibody purification process using three chromatography steps without protein A. J Chromatogr A 2004 1024(1-2), 79-85.

[13] Kuczewski M., Fraud N., Faber R., Zarbis-Papastoitsis G.: Development of a polishing step using a hydrophobic interaction membrane adsorber with a PER.C6-derived recombinant antibody. Biotechnol Bioeng 2010 105(2), 296-305.

[14] Specht R., Han B., Wickramasinghe S. R., Carlson J.O., Czermak P., Wolf A., Reif O. W.: Densonucleosis virus purification by ion exchange membranes. Biotechnol Bioeng 2004 88(4), 465-473.

[15] Gottschalk U., Fischer-Frühholz S., Reif O.W.: Membrane Adsorbers: A Cutting Edge Process Technology at the Threshold. BioProcess 2004 2(5), 56-65.

References

[16] Lim J. A. C., Sinclair A., Kim D. S., Gottschalk U.: BioProcess Economic Benefits of Single-Use Membrane Chromatography in Polishing. International 2007 5(2), 48-58.

[17] Knudsen H. L., Fahrner R. L., Xu Y., Norling L. A., Blank G. S.: Membrane ion-exchange 2001 chromatography for process-scale antibody purification. J Chromatogr A 3 907(1-2), 145-154.

[18] Gottschalk, U.: Downstream processing of monoclonal antibodies: from high dilution to high purity. BioPharm Int 2005 18(6), 42–58.

[19] Schmidt F.: Recombinant expression systems in the pharmaceutical industry. Appl Microbiol Biotechnol 2004 65(4), 363-372.

[20] Fischer R., Stoger E., Schillberg S., Christou P., Twyman R. M.: Plant-based production of biopharmaceuticals. Curr Opin Plant Biol 2004 7(2), 152-158.

[21] Subramanian G.: Bioseparation and Bioprocessing: Biochromatography, Membrane Separations, Modeling, Validation. Wiley-VCH, Weinheim (Germany), 2007.

[22] Yigzaw Y., Piper R., Tran M., Shukla A. A.: Exploitation of the adsorptive properties of depth filters for host cell protein removal during monoclonal antibody purification. Biotechnol Prog 2006 22(1), 288-96.

[23] Goyal S. M., Hanssen H., Gerba C. P.: Simple method for the concentration of influenza virus from allantoic fluid on microporous filters. Appl Environ Microbiol 1980 39(3), 500-4.

[24] Kusnadi A. R., Hood E. E., Witcher D. R., Howard J. A., Nikolov Z. L.: Production and Purification of Two Recombinant Proteins from Transgenic Corn. Biotechnology Progress 1998 14(1), 149-155.

[25] Vicente T., Peixoto C., Carrondo M. J. T., Alves P. M.: Purification of recombinant baculoviruses for gene therapy using membrane processes. Gene Therapy 2009 16, 766–775.

[26] Van Reis R., Zydney A.: Membrane separations in biotechnology. Curr Opin Biotech 2001 12(2), 208-211.

[27] Eibl R., Eibl D.:Single-Use Technology in Biopharmaceutical Manufacture. Wiley, Hoboken (USA), 2011.

[28] Zhou J. X., Solamo F., Hong T., Shearer M., Tressel T.: Viral Clearance Using Disposable Systems in Monoclonal Antibody Commercial Downstream Processing. Biotechnol Bioeng 2008 100(3), 488-946.

[29] Brower M., Buttke A., Pollard D., Tugcu N.: Working Towards an Integrated Antibody Purification Process. Oral presentation at the IBC's 7th International Biopharmaceutical Manufacturing and Development Summit, San Diego (USA), September 12-14, 2011.

[30] Strauss D. M., Scott Lute S., Tebaykina Z., Frey D. D., Ho C., Blank G.S., Brorson K., Chen Q., Yang B.: Understanding the Mechanism of Virus Removal by Q Sepharose Fast Flow Chromatography During the Purification of CHO-Cell Derived Biotherapeutics. Biotechnol Bioeng 2009 104(2), 371-380.

References

[31] Allen L.: Forum Developing Purification Unit Operations for High Titre Monoclonal Antibody Processes. Oral presentation at the 6th European Downstream Technology Forum, Goettingen (Germany), September 7-8, 2010.

[32] Colaco C., Roser B. J., Sen S.: Method for stabilization of biological substances during drying and subsequent storage and composition thereof. Patent US 5955488, 1999.

[33] Mora J., Sinclair A., Delmdahl N., Gottschalk U.: Disposable Membrane Chromatography: Performance Analysis and Economic Cost Model. Bioprocess Intl 2006 4(6), 38-43.

[34] Zarbis-Papastoitsis G.: High Capacity Purification Schemes Integrated with Extreme Cell Density Process –The PER.C6 Story. Oral presentation at the IBC's Antibody Development & Production, Carslbad (USA), March 3-5, 2009.

[35] Shukla A. A., Jiang C., Ma J., Rubacha M., Flansburg L., Lee S.S.: Demonstration of Robust Host Cell Protein Clearance in Biopharmaceutical Downstream Processes. Biotechnol Prog 2008 24(2), 615-622.

[36] McCurley P.: Comparison of Chromatography Media Behavior in Virus-like Particle Purification Processes. Oral presentation at the Downstream Technology Forum, San Diego (USA), October 14, 2010.

[37] Goding J. W.: Monoclonal Antibodies: Principles and Practice: Principles and Practice - Production and Application of Monoclonal Antibodies in Cell Biology, Biochemistry and Immunology. Academic Press Inc., San Diego (USA), 1986.

[38] Smith G., Bright R., Pushko P., Zhang J., Mahmood K.: Functional influenza virus like particles (VLPS). Patent WO 2007047831, 2007.

[39] Sinclair A.: How to Evaluate the Cost Impact of Using Disposables in Biomanufacturing. Biopharm int 2008 21(6), 26-28.

[40] Hubbuch J., Kula M. R.: Isolation and Purification of Biotechnological Products. J Non-Equili Thermodyn. 2007 32(2), 99-127.

[41] Franke A., Forrer N., Butté A., Cvijetić B., Morbidelli M., Jöhnck M., Schulte M.: Role of the ligand density in cation exchange materials for the purification of proteins. J Chromatogr A 2010 1217(15), 2216-2225.

[42] Shukla A., Hubbard B., Tressel T., Guhan S., Low D.: Downstream Processing of monoclonal antibodies – Application of platform approaches. J Chromatogr B 2007 848(1), 28-39.

[43] De Vocht M.: Viral vaccine processes: Towards intensification and quality. Oral presentation at the 7th European Downstream Technology Forum, Goettingen (Germany), May 24-25, 2011.

[44] Kuczewski M., Schirmer E. B.: A Single-Use Purification Platform for Monoclonal Antibodies. Oral presentation at the IBC's 7th Single-Use Applications for Biopharmaceutical Manufacturing Conference, La Jolla (USA), June 14-16, 2010.

References

[45] Langer E. S.: Downstream Disposables: The Missing Link in the Disposables Chain. Biopharm int 2010 23(8), 22-23.

[46] Morbidelli M.: Continuous chromatography (MCSGP) for the purification of monoclonal antibodies –recent advances. Oral presentation at the BioProduction Conference, Barcelona (Spain), October 27-28, 2010.

[47] Eckermann C., Ebert S., Rubenwolf S., Ambrosius, D.: Process for optimizing chromatographic purification processes for biomolecules. Patent WO 2007144353, 2007.

[48] Arakawa T., Philo J. S., Ejima D., Tsumoto K., Arisaka F.: Aggregation Analysis of Therapeutic Proteins, Part 1, General Aspects and Techniques for Assessment. BioProcess Int 2006 4(10), 42-43.

[49] Wolter T., Richter A.: Assays for Controlling Host Cell Impurities in Biopharmaceuticals. BioProcess Int 2005 2(2), 40-46.

[50] Jin M., Szapiel N., Zhang J., Hickey J., Ghose S.: Profiling of Host Cell Proteins by Two-Dimensional Difference Gel Electrophoresis (2D-DIGE): Implications for Downstream Process Development. Biotechnol Bioeng 2010 105(2), 306-316.

[51] World Health Organization: Acceptability of Cell Substrates for Production of Biologicals. WHO Technical Report Series 747, 1987.

[52] CBER: Points to Consider in the Manufacture and Testing of Monoclonal Antibody Products for Human Use. J Immunother 1997 20(3), 214-43.

[53] Han B., Specht R., Wickramasinghe S. R., Carlson J. O.: Binding Aedes aegypti densonucleosis virus to ion exchange membranes. J Chromatogr A 2005 1092(1), 114-24.

[54] Vicente T., Fáber R., Alves P. M., Carrondo M. J., Mota J. P.: Impact of Ligand Density on the Optimization of Ion-Exchange Membrane Chromatography for Viral Vector Purification. Biotechnol Bioeng 2011 108(6), 1347-1359.

[55] Opitz L., Lehmann S., Reichl U., Wolff M. W.: Sulfated membrane adsorbers for economic pseudo-affinity capture of influenza virus particles. Biotechnol Bioeng 2009 103(6), 1144-1154.

[56] Moroni C., Schumann G.: Are endogenous C-type viruses involved in the immune system. Nature 1977 269(5629), 600 – 601.

[57] Petsch D., Anspach F. B.: Endotoxin removal from protein solutions. J Biotechnol 2000 76(2-3), 97-119.

[58] Anspach F. B.: Endotoxin removal by affinity sorbents. J Biochem Biophys Methods 2001 49(1-3), 665-681.

[59] Magalhães P. O., Lopes A. M., Mazzola P. G., Rangel-Yagui C., Penna T. C., Pessoa A. Jr.: Methods of Endotoxin Removal from Biological Preparations: a Review. J Pharm Pharm Sci 2007 10(3), 388-404.

[60] Steck, S.: Endotoxine und ihre Bedeutung bei R&D Applikation, LPS-Entfernung und -nachweismethoden. BIOforum 2006 6, 34-36.

References

[61] CBER: Guideline on Validation of the Limulus Amebocyte Lysat Test as an Endproduct Test for Human and Animal Parenteral Drugs, Biological Products and Medical Devices. Guideline 1987.

[62] Park J. H., Gold D. R., Spiegelman D. L., Burge H. A., Milton D. K.: House dust endotoxin and wheeze in the first year of life. Am J Respir Crit Care Med 2001 163(2), 322-328.

[63] Petsch D., Beeskow T. C., Anspach F. B., Deckwer W. D.: Membrane adsorbers for selective removal of bacterial endotoxin. J Chromatogr B Biomed Sci Appl 1997 693(1), 79-91.

[64] Moore J. M., Patapoff T. W., Cromwell M. E.: Kinetics and thermodynamics of dimer formation and dissociation for a recombinant humanized monoclonal antibody to vascular endothelial growth factor. Biochemistry 1999 38(42), 13960-13967.

[65] Cromwell M. E., Hilario E., Jacobson F.: Protein aggregation and bioprocessing. AAPS J 2006 8(3), 572-579.

[66] Faude A., Kohne F.: Die richtige Formulierung zum Erhalt der Qualität therapeutischer Proteine. BIOspektrum 2011 17(2), 233-235.

[67] Sadavarte R., Fraud N., Ghosh R.: Feasibility study for the removal of monoclonal antibody aggregates using membrane chromatography. Oral presentation at the 61st Canadian Chemical Engineering Conference, Ontario(Canada), October 23-26, 2011.

[68] Arvinte T.: Overcoming Issues Relating to High Concentration Formulation. Oral presentation at the BioProduction Conference, Barcelona (Spain), October 28, 2010.

[69] Frahm B., Brod H., Langer U.: Improving bioreactor cultivation conditions for sensitive cell lines by dynamic membrane aeration. Cytotechnology 2009 59(1), 17-30.

[70] Rathore N., Rajan R. S.: Current Perspectives on Stability of Protein Drug Products during Formulation, Fill and Finish Operations. Biotechnol Prog 2008 24(3), 504-514.

[71] Chen B. L., Arakawa T., Hsu E., Narhi L. O., Tressel T. J., Chien S. L.: Strategies to suppress aggregation of recombinant keratinocyte growth factor during liquid formulation development. J Pharm Sci 1994 83(12), 1657-1661.

[72] Liu H. F., McCooey B., Duarte T., Myers D. E., Hudson T., Amanullah A., van Reis R., Kelley B. D.: Exploration of overloaded cation exchange chromatography for monoclonal antibody purification. J Chromatogr A 2011 1218(39), 6943-6952.

[73] Vermeer A. W., Norde W.: The thermal stability of immunoglobulin: unfolding and aggregation of a multi-domain protein. Biophys J 2000 78(1), 394-404.

[74] Bee J. S., Randolph T. W., Carpenter J. F., Bishop S. M., Dimitrova M. N.: Effects of surfaces and leachables on the stability of biopharmaceuticals. J Pharm Sci 2011 100(10), 4158–4170.

[75] Bloom J. W., Wong M. F., Mitra G.: Detection and reduction of protein A contamination in immobilized protein A purified monoclonal antibody preparations. Journal of Immunological Methods. J Immunol Methods 1989 117(1), 83-89.

References

[76] Arakawa T., Tsumoto K., Nagase K., Ejima D.: The effects of arginine on protein binding and elution in hydrophobic interaction and ion-exchange chromatography. Protein Expr Purif 2007 54(1), 110-116.

[77] Gebauer K. H., Thömmes J., Kula M.R.: Breakthrough performance of high-capacity membrane adsorbers in protein chromatography. Chem Eng Sci 1997 52(3), 405-419.

[78] Ghosh R.: Protein separation using membrane chromatography: opportunities and challenges. J Chromatogr A 2002 952(1-2), 13-27.

[79] Demmer W., Nussbaumer D.: Large-scale membrane adsorbers. J Chromatogr A 1999 852(1), 73-81.

[80] Zhou J. X., Tressel T.: Basic concepts in Q membrane chromatography for large-scale antibody production. Biotechnol Prog 2006 22(2), 341-349.

[81] Shiosaki A., Goto M., Hirose T.: Frontal analysis of protein adsorption on a membrane adsorber. J Chromatogr A 1994 679(1), 1-9.

[82] Arshady R.: Beaded polymer supports and gels: I. Manufacturing techniques. J Chromatogr A 1991 586(2), 181-197.

[83] McCue J. T., Engel P., Ng A., Macniven R., Thömmes J.: Modeling of protein monomer/aggregate purification and separation using hydrophobic interaction chromatography. Bioprocess Biosyst Eng 2008 31(3), 261-275.

[84] Phillips M., Cormier J., Ferrence J., Dowd C., Kiss R., Lutz H., Carter J.: Performance of a membrane adsorber for trace impurity removal in biotechnology manufacturing. J Chromatogr A 2005 1078(1-2), 74-82.

[85] Nussbaumer D., Demmer W.: Verfahren und Vorrichtung zur adsorptiven Stofftrennung. Patent DE 10236664, 2002.

[86] Ubiera A. R., Carta G., Pabst T. M.: Protein Mass Transfer Kinetics in Ion Exchange Media: Measurements and Interpretations. Chem Eng Techol 2005 28(11), 1252–1264.

[87] Tejeda-Mansir A., Montesinos R. M., Guzmán R.: Mathematical analysis of frontal affinity chromatography in particle and membrane configurations. J Biochem Biophys Methods 2001 49(1-3), 1-28.

[88] Demmer W., Grabosch M., Reif O. W., Nagel W., Donzeau M., Reichelt P.: Reinigung von hochmolekularen Verbindungen mittels Affinitätsmembranchromatographie. Patent DE 102004004043, 2005.

[89] Shirataki H., Sudoh C., Eshima T., Yokoyama Y., Okuyama K.: Evaluation of an anion-exchange hollow-fiber membrane adsorber containing γ-ray grafted glycidyl methacrylate chains. J Chromatogr A 2011 1218(17), 2381-2388.

[90] Josic D., Buchacher A., Jungbauer A.: Monoliths as stationary phases for separation of proteins and polynucleotides and enzymatic conversion. J Chromatogr B Biomed Sci Appl 2001 752(2), 191-205.

References

[91] Hanora A., Savina I., Plieva F. M., Izumrudov V. A., Mattiasson B., Galaev I. Y.: Direct capture of plasmid DNA from non-clarified bacterial lysate using polycationgrafted Monoliths. J Biotechnol 2006 123(3), 343-55.

[92] Cox M. M., Phillips G. N.: The Handbook of Proteins: Structure, Function and Methods. Wiley-VCH, Weinheim (Germany), 2007.

[93] Brorson K., Shen H., Lute S., Pérez J. S., Frey D. D.: Characterization and purification of bacteriophages using chromatofocusing. J Chromatogr A 2008 1207(1-2), 110-121.

[94] Woo M., Khan N. Z., Royce J., Mehta U., Gagnon B., Ramaswamy S., Soice N., Morelli M., Cheng K. S.: A novel primary amine-based anion exchange membrane adsorber. J Chromatogr A 2011 1218(32), 5386-5392.

[95] Riordan W., Heilmann S., Brorson K., Seshadri K., He Y., Etzel M. R.: Design of salt-tolerant membrane adsorbers for viral clearance. Biotechnol Bioeng 2009 103(5), 920-929.

[96] Malone T., Li M.: PAT-Based In-Line Buffer Dilution. BioProcess Int 2010 8(1), 40-49.

[97] Juhnke M.: Improving Processes and Delivering Business Benefits through QbD: Milling of an API. Oral presentation at the IQPC/Pharma IQ's Annual PAT and Quality by Design, London (Great Britain), February 24-25, 2009.

[98] Taticek R.: Update on the implementation of QbD at Genentech and Participation in the FDA QbD Pilot Program. Oral presentation at the BioProduction Conference, Barcelona (Spain), October 27-28, 2010.

[99] Hardin A. M., Harinaraya C., Malmquist G., Axén A., van Reis R.: Ion exchange chromatography of monoclonal antibodies: Effect of resin ligand density on dynamic binding capacity. J Chromatogr A 2009 1216(20), 4366–4371.

[100] Mannuzza F. J., Montalto J. G.: Is Bovine Albumin Too Complex to Be Just a Commodity. BioProcess Int 2010 8(2), 40–43.

[101] Hancock W. S., Sparrow J. T.: Use of mixed-mode, high-performance liquid chromatography for the separation of peptide and protein mixtures. J Chromatogr 1981 206(1), 71-82.

[102] Avramescu M. E., Borneman Z., Wessling M.: Mixed-matrix membrane adsorbers for protein separation. J Chromatogr 2003 1006(1-2), 171-183.

[103] Gallop M. A., Barret R. W., Dower W. J., Fodor S. P. A., Gordon M.: Applications of combinatorial libraries to drug discovery. 1. Background and peptide combinatorial libraries. J Med Chem 1994 37(9), 1233–1251.

[104] Britsch L.: Parallel-Chromatographie im Miniaturmaßstab. GIT Separation 2005 2, 18–19.

[105] Teeters M., Bezila D., Alred P., Velayudhan A.: Development and application of an automated, low-volume chromatography system for resin and condition screening. Biotechnol J 2008 3(9-10), 1212-1223.

References

[106] Dieterle M., Studts J. M., Rathjen T., Wenzel D., Ambrosius D.: The Challenges of Scale-Up and High-Throughput Methods in Downstream Development. Oral presentation at the High-Throughput Process Development Conference, Krakow (Poland), October 4-7, 2010.

[107] Rodrigo G., Nilsson-Välimaa K.:Developing a MAb Aggregate Removal Step by High Throughput Process Development. BioPharm Int 2010 23(4), 4-6.

[108] Bensch M., Schulze Wierling P., von Lieres E., Hubbuch J.: High Throughput Screening of Chromatographic Phases for Rapid Process Development. Chem Eng Technol 2005 28(11), 1274-1284.

[109] Schulze Wierling P., Bogumil R., Knieps-Grünhagen E., Hubbuch J.: High-throughput screening of packed-bed chromatography coupled with SELDI-TOF MS analysis: monoclonal antibodies versus host cell protein. Biotechnol Bioeng 2007 98(2), 440-450.

[110] Brown A., Bill J., Tully T., Radhamohan A., Dowd C.: Overloading ion-exchange membranes as a purification step for monoclonal antibodies. Biotechnol Appl Biochem 2010 56(2), 59-70.

[111] Mora J., Dolan S., Fraud N.: Scaling Up Disposable Membrane Chromatography. BioPharm Int 2006.

[112] Rege K., Pepsin M., Falcon B., Steele L., Heng M.: High-throughput process development for recombinant protein purification. Biotechnol Bioeng 2006 93(4), 618-630.

[113] Kökpinar O., Harkensee D., Kasper C., Scheper T., Zeidler R., Reif O. W., Ulber R.: Innovative modular membrane adsorber system for high-throughput downstream screening for protein purification. Biotechnol Prog 2006 22(4), 1215-1219.

[114] Harkensee D., Kökpinar Ö., Walter J., Kasper C., Beutel S., Reif O. W., Scheper T. Ulber R.: Fast Screening for the Purification of Proteins Using Membrane Adsorber Technology. Eng Life Sci 2007 7(4), 388 – 394.

[115] Kramarczyk J. F., Kelley B. D., Coffman J. L.: High-throughput screening of chromatographic separations: II. Hydrophobic interaction. Biotechnol Bioeng 2008 100(4), 707-720.

[116] Kelley B. D., Switzer M., Bastek P., Kramarczyk J. F., Molnar K., Yu T., Coffman J.: High-throughput screening of chromatographic separations: IV. Ion-exchange. Biotechnol Bioeng 2008 100(5), 950-963.

[117] Bergander T., Nilsson-Välimaa K., Oberg K., Lacki K. M.: High-throughput process development: determination of dynamic binding capacity using microtiter filter plates filled with chromatography resin. Biotechnol Prog 2008 24(3), 632-639.

[118] Studts J., Rathjen T., Wenzel D., Stolzenberger J., Ambrosius D.: Impact of automation in process optimization on downstream development timelines. Oral presentation at the BioProcess international, Vienna (Austria), May 18-21, 2010.

References

[119] Leuthold M.: High Throughput Downstream Screening System for protein purification using membrane adsorbers. Oral presentation at the High-Throughput Process Development Conference, Krakow (Poland), October 4-7, 2010.

[120] Fiske D. H., Subbarow Y.: The Colorimetric Determination of Phosphorus. J Biol Chem 1925 66, 375–400.

[121] News from industry. Biotechnology Journal 2006 1(1), 13-20.

[122] Faber R., Yang Y., Gottschalk U.: Salt tolerant interaction chromatography for large-scale polishing with convective media. BioPharm Int Suppl 2009, 11-14.

[123] Demmer W., Faber R., Hörl H. H., Kiss C., Nussbaumer D.: Cellulose hydrate membrane, method for the production thereof, and use thereof. Patent US 20110147292, 2011.

[124] Kern G., Krishnan M.: Virus removal by Filtration: Points to consider. BioPharm Int 2006 19(10), 32-41.

[125] Cipriano D.: Enhancing viral clearance performance: lessons from data mining and improving virus clearance. Oral presentation at the 5th European Downstream Technology Forum, Goettingen (Germany), June 16-17, 2009.

[126] Etzel M. R., Riordan M.: Clearance of biological impurities using improved membrane adsorbers. Oral presentation at the 236th ACS National Meeting, Philadelphia (USA), August 17-21, 2008.

[127] Masel R. I.: Principles of Adsorption and Reaction on Solid Surfaces. Wiley, Hoboken (USA), 1996.

[128] Wedler G.: Adsorption – Eine Einführung in die Physisorption und die Chemisorption. Verlag Chemie, Heidelberg (Germany), 1970.

[129] Carta G., Jungbauer A.: Protein Chromatography: Process Development and Scale-Up. Wiley-VCH, Weinheim (Germany), 2010.

[130] Jungbauer A.: Adsorption isotherms in protein chromatography Combined influence of protein and salt concentration on adsorption isotherm. J Cromatogr A 1996 734(1), 183-194.

[131] Yang H., Viera C., Fischer J., Etzel M. R.: Purification of a Large Protein Using Ion-Exchange Membranes. Ind Eng Chem Res 2002 41(6), 1597–1602.

[132] Faber R., Leuthold M.: Multi-well plate having filter medium and use thereof. Patent WO 2011113508, 2011.

[133] Glynn J., Chen B.: Effect of Variations of pH and Salt Concentration on Purification Efficiency for Q Membrane. Oral presentation at the Downstream Technology Forum, San Diego (USA), October 14, 2010.

[134] McAlister M., Aranha H., Larson R.: Use of bacteriophages as surrogates for mammalian viruses. Dev Biol 2004 118, 89-98.

[135] Syngouna V. I., Chrysikopoulus C. V.: Interaction between Viruses and Clays in Static and Dynamic Batch Systems. Environ Sci Technol 2010 44, 4539-4544.

References

[136] Eiserling F., Pushkin A., Gingery M., Bertani G.: Bacteriophage-like particles associated with the gene transfer agent of Methanococcus voltae PS. J Gen Virol 1999 80 (12), 3305-3308.

[137] Michen B., Graule T.: Isoelectric points of viruses. J Appl Microbiol 2010 109(2), 388-397.

[138] Jin M., Szapiel N., Zhang J., Hickey J., Ghose S.: Profiling of host cell proteins by two-dimensional difference gel electrophoresis (2D-DIGE): Implications for downstream process development. Biotechnol Bioeng 2010 105(2), 306-316.

[139] Kraetzig M.: Untersuchung zur Entfernung von Host cell Proteinen mit Membranadsorbern. Master thesis, Institute of Technical Chemistry, University Hannover, 2011.

[140] Mürer E. H., Levin J., Holme R.: Isolation and studies of the granules of the amebocytes of Limulus polyphemus, the horseshoe crab. J Cell Physiol 1975 86, 533-542.

[141] Kessler W.: Multivariate Datenanalyse: für die Pharma-, Bio- und Prozessanalytik. Wiley-VCH, Weinheim (Germany), 2006.

[142] Kessler R. W.: Strategien und Fallbeispiele aus der industriellen Praxis. Wiley-VCH, Weinheim (Germany), 2006.

[143] Deparis V.: Applying enhanced quality of design approach for the manufacture of a Fc-fusion protein. Oral presentation at the BioProduction Conference, Barcelona (Spain), October 27-28, 2010.

[144] Sonderegger C.: Target-directed process development & characterisation. Oral presentation at the BioProduction Conference, Barcelona (Spain), October 27-28, 2010.

[145] Multivariate calibration: What is in chemometrics for the analytical chemist. Analytica Chimica Acta 2003 500(1-2), 185-194.

[146] Kneen M., Farinas J., Li Y., Verkman A. S.: Green Fluorescent Protein as a Noninvasive Intracellular pH Indicator. Biophys J 1998 74(3), 1591-1599.

[147] Uetz P., Pohl E.: Protein-Protein- und Protein-DNA Interaktionen. In: Wink et al., Molekulare Biotechnologie. Wiley-VCH, Weinheim (Germany), 2004.

[148] Sata T., Yamaguchi T., Matsusaki K.: Effect of Hydrophobicity of Ion Exchange Groups of Anion Exchange Membranes on Permselectivity between Two Anions. J Phys Chem 1995 99(34), 12875–12882.

[149] Zhang F., Skoda M. W., Jacobs R. M., Zorn S., Martin R. A., Martin C. M., Clark G. F., Weggler S., Hildebrandt A., Kohlbacher O., Schreiber F.: Reentrant Condensation of Proteins in Solution Induced by Multivalent Counterions. Phys Rev Lett 2008 101(14), 148101.

[150] Holmes-Farley S. R., Mandeville W. H., Ward J., Miller K. L.: Design and characterization of sevelamer hydrochloride: a novel phosphate- binding pharmaceutical. J Macromol Sci Pure Appl Chem 1999 36(7-8), 1085-1091.

References

[151] Kang Y. K., Ng S., Lee J., Adaelu J., Qi B., Persaud K., Ludwig D., Balderes P.: Development of an Alternative Monoclonal Antibody Polishing Step. BioPharm Int 2012 25(5), 34-46.

[152] Wang L., Hale G., Ghosh R.: Non-size-based membrane chromatographic separation and analysis of monoclonal antibody aggregates. Anal Chem 2006 78(19), 6863-6867.

[153] Xia F., Nagrath D., Garde S., Cramer S. M.: Evaluation of selectivity changes in HIC systems using a preferential interaction based analysis. Biotechnol Bioeng 2004 87(3), 354-363.

[154] He F., Phan D. H., Hogan S., Bailey R., Becker G. W., Narhi L. O., Razinkov V. I.: Detection of IgG aggregation by a high throughput method based on extrinsic fluorescence. J Pharm Sci 2010 99(6), 2598-2608.

[155] Juditsky A., Hjalmarsson H., Benveniste A., Delyon B., Ljung L, SjÖberg J., Zhang Q.: Nonlinear black-boxmodels in system identification: Mathematical foundations. Automatica 1995 31(12), 1725-1750.

[156] Faber R., Leuthold M.: Verfahren zur Abtrennung von Viren aus einem Kontaminanten enthaltenen flüssigen Medium. Patent DE 102010056817, 2012.

[157] Po H. N., Senozan N. M.: The Henderson-Hasselbalch Equation: Its History and Limitations. J Chem Educ 2001 78(11), 1499-1503.

[158] Chang R.: Physical Chemistry for the Chemical and Biological Sciences. University Science Books, Mill Vally (USA), 2000.

[159] Delaunay B. N.: Sur la sphère vide. In: Bulletin of Academy of Sciences of the USSR 7. 1934 6, 793-800.

[160] Yang T. T., Cheng L., Kain S. R.: Optimized codon usage and chromophore mutations provide enhanced sensitivity with the green fluorescent protein. Nucleic Acids Res 1996 24, 4592–4593.

[161] Böttner M.: Die Expression humaner Proteine in der Hefe Pichia pastoris: Hochdurchsatzverfahren und bioinformatische Identifizierung von Expressionbeeinflussenden Sequenzmerkmalen. Dissertation, Department of Process Science, University Berlin, 2004.

[162] Otto M.: Analytische Chemie. Wiley-VCH, Weinheim (Germany), 2011.

[163] Lindsay G. K., Roslansky P. F., Novitsky T. J.: Single-step, chromogenic Limulus amebocyte lysate assay for endotoxin. J Clin Microbiol 1989 27(5), 947-951.

8 Appendix

8.1 Materials

8.1.1 Chemicals and biomolecules

The following chemicals and biomolecules were used for buffer preparation, screening studies and detection assays.

Table 6: Chemicals.

Product	Manufacturer
CHES	Sigma Corporation, St. Louis, USA
Tris	Roth, Karlsruhe, Germany
BIS-Tris	Roth, Karlsruhe, Germany
Acetic acid	Roth, Karlsruhe, Germany
Citric acid	Merck, Darmstadt, Germany
Sodium acetate	Roth, Karlsruhe, Germany
Disodium phosphate	Merck, Darmstadt, Germany
Monosodium phosphate	Merck, Darmstadt, Germany
Sodium chloride	Roth, Karlsruhe, Germany
Hydrogen chloride	Roth, Karlsruhe, Germany
Sodium hydroxide	Fluka Chemie AG
Decon 90	Decon Laboratories Limited, United Kingdom
Ascorbic acid	Roth, Karlsruhe, Germany

Appendix

Product	Supplier
Acid sulphur	Roth, Karlsruhe, Germany
Ammonium molybdate	Roth, Karlsruhe, Germany
Beta-Glucan-Blocker Kit	Lonza, Basel, Switzerland
Endotoxin, E. coli O55:B5	Lonza, Basel, Switzerland
Nutrient Broth	BD, Franklin Lakes, USA
Nutrient Broth Agar	BD, Franklin Lakes, USA
Acetone	Roth, Karlsruhe, Germany
Ethanol, 96 %	VWR International GmbH
BCA-Assay Kit A&B	Pierce, Rockford, USA
PicoGreen®	Invitrogen AG, Carlsbad, USA
Endochrome-K	Charles River
Trypticase Soy Agar	BD, Franklin Lakes, USA
Trypticase Soy Broth	BD, Franklin Lakes, USA

A 1 % solution of the detergent Decon 90 was used for cleaning the pipetting tips of the robotic platform prior to each pipetting step using different buffers or during sampling.

Appendix

Table 7: Biomolecules.

Substance	Supplier
BSA	Kraeber, Ellerbek, Germany
Salmon sperm DNA	Biomol GmbH, Hamburg, Germany
Calf thymus DNA	Sigma, Deisenhofen, Germany
Endotoxin	Lonza Group Ltd, Basel, Switzerland

The GFP was produced by the cultivation of *Pichia pastoris* and Bacteriophage ΦX174 by the cultivation of *E.coli*. The purified Bacteriophage ΦX174 and cell-free GFP containing solutions were kindly donated by the R&D department at Sartorius-Stedim Biotech GmbH. For more details of the cultivation and components of the culture media see Appendix 8.1.5 and 8.1.6.

8.1.2 Buffers and buffer preparation

Stock solutions were used for the preparation of the chromatographic media. Each stock solution was filtered through 0.2 µm. The pH was adjusted at room temperature. Table 8 contains the buffers used.

Appendix

Table 8: Basic buffers. The conductivities were determined for a buffer concentration of 20 mM.

pH range	Buffer	Conductivity
4.0 – 5.5	Acetate	0.4 – 0.9 mS/cm
6.0 – 7.0	BIS-Tris	0.5 – 1.6 mS/cm
7.5 – 9.0	Tris	0.5 – 1.8 mS/cm
9.5 – 10.0	CHES	0.4 – 1.4 mS/cm

Two steps were necessary for the automatic preparation of buffers using the robotic platform. The calibration of the buffer system was done first. Acid was titrated with base depending on the buffer system. The amount of base and acid necessary to adjust a specific pH of a buffer was determined. 16 solutions were prepared, each with a different proportion of acid or base. The target molarity of each buffer was 20 mM. Three different types of buffer were prepared:

- Acetate: the acid was titrated against the corresponding base
- Tris, BIS-Tris: the base was kept constant and the amount of acid was varied
- CHES: the acid was kept constant and the amount of base was varied

The determined functions correspond to the Henderson-Hasselbalch equation [157]. The buffers were only used in the pH range described in Table 8.

Appendix

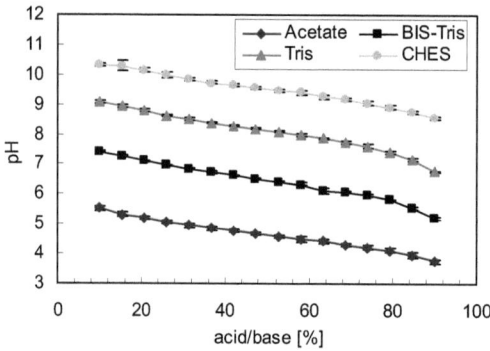

Figure 41: Adjustment of pH for the different buffers. The pH was adjusted by the addition of base or acid. The percentage of acid or base refers to the concentration of CHES, Tris and BIS-Tris. In the case of acetate the percentage of acid refers to the total amount of the buffer. The calibrations were done at five different sodium chloride concentrations (0, 30, 50, 100 and 150 mM/l). The average pH values are shown. The deviation of the pH was ≤ 0.16.

The performance using ion exchange chromatography in the downstream purification process depends inter alia on the conductivity of the feed solution. Sodium chloride was added in for adjustment of different conductivities. Figure 42 shows the conductivity with increasing concentration of sodium chloride.

Appendix

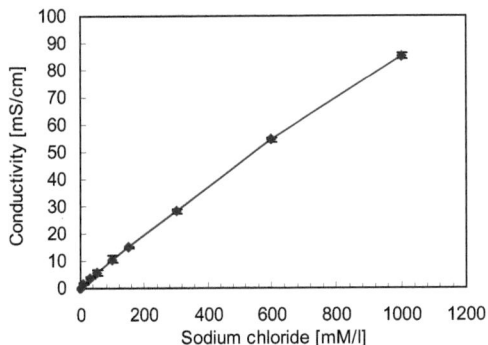

Figure 42: Conductivity depending on sodium chloride concentration. The dependence of conductivity from the sodium chloride concentrations assumed to be linear. The conductivities were determined of 12 buffer preparations.

The calibration data were stored in an Excel spreadsheet. The amount of acid or base needed to adjust the required pH was calculated by linear regression in order. The percentages of further components like protein, salt and water were determined next. Furthermore, the total volume of each buffer and the allocation of stock reservoirs were saved in the file. Reading the receipt from the Excel file, the robotic software calculates the piston stroke for transport and mixing the corresponding volumes. Based on the concentrations of the stock solutions and the percentage of the buffer any buffer composition could be mixed.

According to Debye-Hueckel the salt concentration have influences on the activity of protons as affected by the addition of chloride ions [158]. An increasing sodium chloride concentration could cause a change to the pH during buffer preparation. The effect has not been considered in this thesis. The deviation of the pH value was < 0.2 during the experiments. An extended calibration protocol could also address this effect.

Phosphate and citrate stock solutions for examining the influence of multivalent salt ions were prepared. The pH of the stock solution corresponds to the pH of

Appendix

the final buffer for the experiment. The multivalent salt was considered as a buffer component.

8.1.3 Equipment

The following equipment was used for experimental setup

Table 9: Equipment.

Equipment	Supplier
Sartobind STIC® Membrane (IEX2)	Sartorius Stedim Biotech GmbH, Goettingen, Germany
Sartobind Q Membrane (IEX1)	Sartorius Stedim Biotech GmbH, Goettingen, Germany
Lissy 2002, modular automated liquid handling system, software Runner Version 7.1.7	Zinsser Analytic GmbH, Frankfurt, Germany
Tecan XSafire plate reader	Tecan Group AG, Maennedorf, Switzerland
Valves	Festo, Esslingen, Germany
Pressure sensor VAM-40-V1/0-R1/8-EN	Festo, Esslingen, Germany
FLx800 Fluorescence Microplate Reader	BioTek Instruments INC, Winooski, USA
HPLC Dionex, pump system P580, sampler ASI-100, detector 170S/340S, Thermostat STH 585, Software Chromeleon v 6.8	Dionex Corporation, Sunnywale, USA
TSK 3000SWxl	Tosoh Bioscienc, Stuttgart, Germany
Sartobind Q MA75	Sartorius Stedim Biotech GmbH, Goettingen, Germany
Vacuum filtration unit	Sartorius Stedim Biotech GmbH, Goettingen, Germany

Appendix

Equipment	Supplier
96-well micro titer plate UV-Star flat	Greiner, Kremsmuenster, Austria
Petri dishes, 90mm	Greiner, Kremsmuenster, Austria
Masterblock 96-well 2 ml	Greiner, Kremsmuenster, Austria
Vacuum pump 16612	Sartorius Stedim Biotech GmbH, Goettingen, Germany
Arium G11 VF	Sartorius Stedim Biotech GmbH, Goettingen, Germany
Multipipette plus	Eppendorf AG, Hamburg, Germany
ÄKTA prime	Amersham Biosciences, Uppsala, Sweden
Autoclave, Varioklave	HP, Oberschleissheim, Germany
Heating block, Thermomixer	Eppendorf AG, Hamburg, Germany
Falcon tubes	Greiner, Kremsmuenster, Austria
Polystyrene tubes	Roth, Karlsruhe, Germany
Analytic balance Genius	Sartorius Stedim Biotech GmbH, Goettingen, Germany
Photometer Genesys 10UV	Thermo Fischer Scientific, Waltham, USA
Shaker Vortex	Gemmy industrial, MinChuan, Taiwan
Pipette tip	Sarstedt AG & Co, Nuernberg, Germany
Sterile filter Minisart 0.2µm	Sartorius Stedim Biotech GmbH, Goettingen, Germany
96-well half area plates	Greiner, Kremsmuenster, Austria
Water quench 3042	Koettermann, Uetze, Germany

Appendix

For further equipment for the cultivation to prepare GFP and Bacteriophage ΦX174 is described in Appendix 8.1.5 and 8.1.6.

8.1.4 Software

To program the working sequences on the robot and to control individual peripherals the software ZA runner from Zinsser Analytic (Frankfurt, Germany) was used. The assignment of each pipetting step and buffer preparation was imported from Excel. For the calculation of buffers Microsoft Office Excel Solver tool was used. The reader server RdrOle4 controlled the Tecan plate reader and the measurement procedures. The evaluation of chromatograms was performed using the software Chromeleon 6.8 for the size exclusion chromatography and Unicorn for the FPLC.

The data analysis and visualization was carried out using the program MATLAB® (The MathWorks, Inc, Natick, United States of America). The generation of contour plots was performed with the experimental data. Three basic commands have been used.

- 'Meshgrid': In the experimental space, described by the influencing factors (for example, salt concentration and pH), a uniform matrix was created. The matrix was limited to the minimum and maximum value of each influencing parameter. The final resolution of the regression of influencing factors was set for 10 sections.

- 'Griddata': This command was used to linearly interpolate the intermediate values specified by the uniform matrix to the experimental data. This method is based on a Delaunay triangulation of the data [159].

- 'Contour': This was the actual visualization. A contour plot of the dependent variable (for example, binding or LRV) was generated, where the dependent variable displayed as color with respect to the influencing factors.

8.1.5 Preparation of Green Fluorescent Protein

The fermentation of *Pichia pastoris X-33 eGFP* was performed using a BIOSTAT® Cplus bioreactor (Sartorius Stedim Biotech GmbH, Goettingen, Germany). Enhanced GFP (eGFP) shows a higher luminosity and allows the use of mammalian cells [160]. The used composition of the nutrient solution was given by the thesis of Boettner [161] (conductivity 6 mS/cm, pH 7). The carbon source was 20 g/l glucose. After 40 h of cultivation 2.8 ml/(h*l) methanol was added for 24 h to initiate the production of GFP. The total fluid volume of the cultivation was 7.5 l. The fermentation was terminated after another 24 h. The fermentation broth was sedimented for 72 h at 4 °C. Two liter of supernatant were purified using a cross-flow system (Vivaflow 200 0.2 µm PES and Vivaflow 50 100K RC, Sartorius Stedim Biotech GmbH, Goettingen, Germany). The volume of the concentrated solution was 190 ml. Furthermore, a chromatographic step was used to separate further components. The solution was loaded to an anion exchanger membrane adsorber (Sartobind Q MA75, Sartorius Stedim Biotech GmbH, Goettingen, Germany). After washing with 40 ml buffer (20 mM Tris pH 8), a linear gradient elution (40 ml, 0 to 1 M sodium chloride in 20 mM Tris buffer pH 8) was used to collect several fractions of 5 ml. Each fraction was analyzed by size exclusion chromatography and fluorescence was measured. The buffers for size exclusion chromatography were used according to the recommended instructions of the manufacturer of the column. The fraction which showed the largest fluorescence as well as the expected GFP-specific peak was portioned frozen at -40 C°.

Appendix

The protein concentration of the fraction of the highest fluorescence was determined by the bicinchoninic (BCA) assay. A reagent kit (PIERCE, Rockford, IL, USA) was used according to the instructions. The BSA equivalent concentration of the GFP stock solution was 1.6 mg/ml.

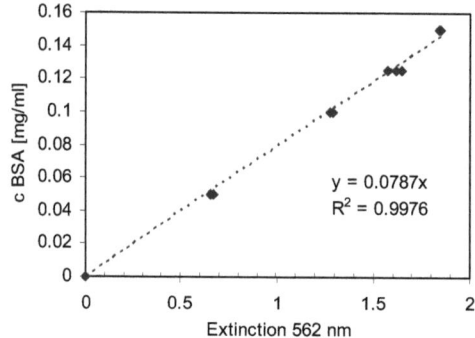

Figure 43: Determination of protein concentration using BCA. A dilution series of BSA was used to create a calibration. The linear calibration function was used to calculate the BSA equivalent protein concentration of the GFP. Before measurement the GFP stock solution was 30-fold diluted (extinction 0.68).

8.1.6 Preparation of Bacteriophage ΦX174

For the spiking studies a concentrated phage stock solution was prepared. The Bacteriophage Φ X174 (ATCC 13706-B1) was produced by a fermentation of *Escherichia coli C* (ATCC 13706). In a disposable bioreactor (Biostat® 50L CultiBag RM, Sartorius Stedim Biotech GmbH, Goettingen, Germany) *E.coli* was grown. The fermentation volume was 20 l; the media was 30 g/l Trypticase Soy broth (Sigma, Deisenhofen, Germany). After a certain cell density was achieved, the bacterial culture was inoculated with phages. The phage production continued for 4 h. Subsequently, the reaction was stopped by filtration. Cell harvest and sterile filtration were done using a depth filter

Appendix

cascade (Sartopure PP2 MidiCap 8 – 5 µm and Sartopore 2XLG 0.8 – 0.2 µm, Sartorius Stedim Biotech, Goettingen, Germany). By cross-flow filtration (30 kDa Sartocon Slice Cassette, Sartorius Stedim Biotech, Goettingen, Germany) the permeate was reduced to a final volume of 600 ml. The next step was a precipitation. The solution was mixed with a polyethylene glycol solution of 30 %. For the separation the mixture was centrifuged at 8430 rpm for 90 min (Sorvall RC-6 centrifuge, Thermo Fisher Scientific, Bonn, Germany). The pellets were resuspended with 25 ml of the storage buffer and stored at -70 °C. The phage concentration of the stock solution was approximately 1×10^{10} PFU/ml. Furthermore, the stock solution was analyzed using the PicoGreen assay. A concentration of 1.1 µg/ml equivalent to salmon sperm DNA was determined. Because of the high dilution for phage spiking studies, no interferences were expected from the DNA. Thus, during the experiments using a phage spike, the DNA concentration was below the detection limit of the phage stock solution.

Appendix

8.2 Methods

The methods used are described in the following. As mentioned, detection assays are often influenced by the buffer conditions. Unless otherwise stated, during this thesis it was assumed that the effect of specific buffer conditions is equal for the initial feed solution and flow through or elution sample. During the screening, the initial solutions containing the target molecules were prepared from the same stock solution. During the experiments the detection method was performed at standard buffer conditions to determine the initial concentration. Published data and the measurements of initial solutions indicated the buffer conditions where the detection was still possible. This considerably reduced the required efforts during the analysis. Thus, the breakthrough or LRV could be determined.

8.2.1 Photometric determination of protein and DNA

The detection of BSA and DNA concentrations was performed by a photometric determination based on the Lambert-Beer law. The extinctions were measured at 280 nm for protein and 260 nm for DNA.

Appendix

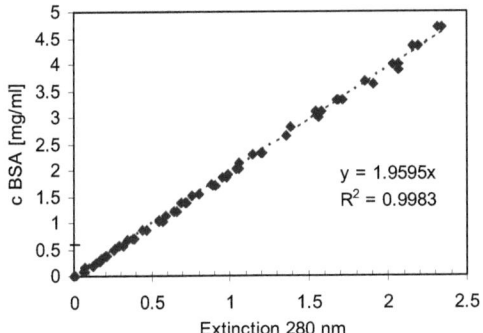

Figure 44: Calibration of BSA using the Tecan plate reader at 280 nm. 24 different concentrations were prepared using the automated liquid handling system. Then two samples of each falcon tube were transferred into the 96-well plate. The buffer was 20 mM Tris at pH 7.4. The calibration was repeated three times. A standard procedure was implemented using the Zinsser control software to repeat the calibration several times. The limit of quantification was set at 0.1 mg/ml.

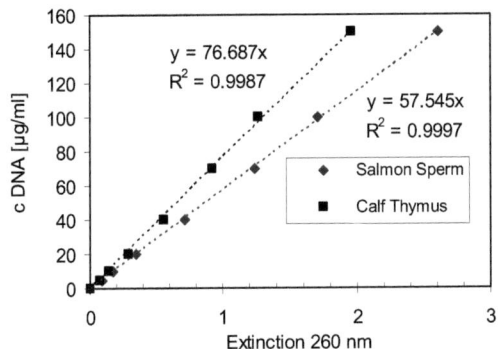

Figure 45: Calibration of DNA using the Tecan plate reader at 260 nm. 8 different concentrations were prepared using the automated liquid handling system. Then two samples of each falcon tube were transferred into the 96-well plate. The buffer was 20 mM Tris at pH 7.4. The limit of quantification was set at 1 µg/ml.

Appendix

8.2.2 Phosphate detection

To exclude a cross-contamination of buffer components in adjacent wells of the 96-well holder the detection method of orthophosphate based on the formation of phosphomolybdate blue was used. All used equipment was cleaned using distilled water. Ammonium molybdate, sulphuric acid and ascorbic acid were mixed in an aqueous solution. Phosphomolybdatic acid is formed in an acidic environment with the presence of phosphate. Ascorbic acid reduces molybdate to phosphomolybdate blue. The detection reagent consists of 50 ml 100 g/l ascorbic acid, 50 ml 6 M sulphuric acid, 50 ml 25 g/l ammonium molybdate and 100 ml of distilled water. 500 µl of the reagent was mixed with 500 µl of sample solution in a 96-well collection plate. The mixture was shaken and heated for 10 min at 70 °C. 300 µm of each solution was transferred to a 96-well microtiter plate and measured at 820 nm using the Tecan plate reader. The quantification limit was defined as three times the blank value.

8.2.3 GFP assay

GFP was detected by the measurement of fluorescence using the Tecan plate reader. The excitation wavelength was 475 nm and the emission wavelength 509 nm. The bandwidth of excitation and emission was 12 nm. Because lowest possible GFP concentrations were intended be used, the detection limit of GFP was evaluated.

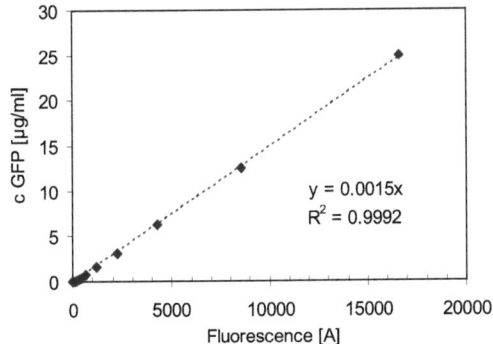

Figure 46: Fluorescence of GFP depending on the concentration. The GFP stock solution was diluted. The lowest concentration was 0.05 µg/ml. Buffer was 20 mM Tris at pH 8. Four samples were measured for each concentration. The gain was automatically determined by the plate reader. The maximum coefficient of variation was 10 % at a concentration of 0.05 µg/ml.

The quantification limit was set at 0.15 µg/ml. During the screening experiments, the maximum GFP concentration was 1.6 µg/ml. In this concentration range the gain was fixed at 150 for all measurements to ensure that the measurement scale was not exceeded.

The fluorescence of GFP is influenced by the pH. The quantification of the fluorescence was performed using the fluorescent area.

Appendix

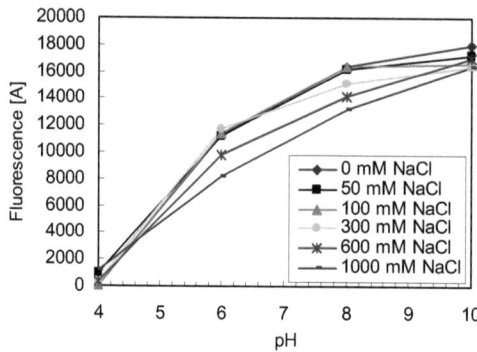

Figure 47: Florescence of GFP depending on pH and sodium chloride. The concentration of GFP was 0.8 µg/ml; this corresponded to 0.05 % of the GFP stock solution. Buffers were 20 mM sodium acetate at pH 4, 20 mM BIS-Tris at pH 6, 20 mM Tris at pH 8 and 20 mM CHES at pH 10. For each buffer condition two samples were measured.

Sufficient fluorescence for quantification was detected in the pH range from 6 to pH 10. The concentration of sodium chloride showed only little influence.

8.2.4 Phage assay

An infectivity assay was used to determine the concentration of Bacteriophage ΦX174. The method based on the same host cell organism (*E. coli*) as for the preparation of phages. Agar plates with nutrient medium were prepared. The bacteria are plated out; during the incubation *E. coli* cells will grow rapidly and form dense opaque bacterial lawns on the surface of the plate. The presence of infectious phage ΦX174 causes lysis of host cells. Thereby, clear areas, called plaques, were formed in the bacterial lawn due to bacterial lysis. A single plaque was defined as a plaque-forming unit (PFU). If necessary, samples were diluted several times in steps of 1:10 until it was possible to determine the numbers of plaques clearly separate from each other.

Appendix

To calculate removal, the phage titer was determined for each initial solution and each sample at various step dilutions. 150 µl of the *E.coli* suspension and sample were mixed with top agar (1.3 % Trypticase Soy Agar BD 211043). Then the mixture was plated on agar (4 % Trypticase Soy Agar BD 211043 in). After incubation at 37 °C for 24 hours, plaque-forming units were counted. The titer was calculated by

$$Titer = \frac{P}{E} x \frac{D}{V_{Sample}} \qquad (8.1)$$

Where P was the number of countable plaques, E the sum of weighting, D the lowest evaluated dilution and V_{Sample} the volume of the sample. Starting from the first countable dilution with E = 1, the weighting decreases by a factor of 10 for each further dilution.

When no plaque was found after incubation, the detection limit of the phage assay was used. The probability of the detection of phages was assumed to be 95 %. According to the relevant guidelines the detection limit c_{lim} was calculated at approximately 20 PFU/ml.

$$c_{lim} = \frac{\ln(0.05)}{V_{Sample}} \qquad (8.2)$$

For each sample a parallel determination of the phage titer was performed. To evaluate the deviation of the LRV the coefficient of variation of phage titer was determined. For three different phage concentrations the standard deviation was calculated. Assuming that the first concentration in Table 10 was the spiking solution, the deviation of LRV was calculated with respect to the lower phage levels in Table 10 using the quadratic error propagation.

Appendix

Table 10: Deviation of ΦX174 plaque assay.

c [PFU/ml]	Standard deviation [PFU/ml]	Coefficient of variation [%]	LRV
1.50E+07	2.93E+06	19.5	-
5.99E+03	1.90E+03	31.8	3.14 ±0.24
5.86E+01	3.43E+01	58.5	5.27 ±0.47

8.2.5 DNA assay

The quantification of low-concentrated salmon sperm and calf thymus DNA was performed using the Quant-iT™ PicoGreen® assay according to the manufacturer's instructions. For this a 10-fold TE stock buffer (100 mM Tris pH 7.5 and 10 mM EDTA) was prepared. The PicoGreen dsDNA reagent was diluted 1 to 200 by mixing 20 µl reagents with 3980 µl using TE-buffer. During the calibration for each type of DNA a stock solution of 2 mg/ml was prepared in TE-buffer and diluted in three steps to 2 µg/ml. In further dilution steps, using TE-buffer, several concentrations with a total volume of 100 µl were prepared for the standard curve. Each DNA solution was mixed with 100 µl PicoGreen® reagent. The mixture was shaken for 3 min and measured using the Tecan plate reader (excitation wavelength 480 nm, emission wavelength 520 nm, gain 100).

Appendix

Table 11: Protocol for preparing PicoGreen® standard curve.

2µg/ml DNA [µl]	TE-Puffer [µl]	c DNA [ng/ml]
0	100	0
0.1	99.9	0.2
1	99	2
5	95	10
10	90	20
25	75	50
50	50	100
75	25	150
90	10	180
100	0	200

Appendix

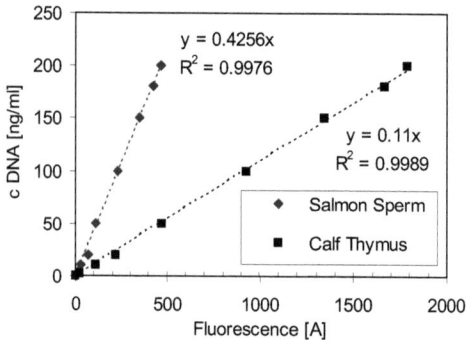

Figure 48: Standard curve of DNA using the PicoGreen® assay. The limit of quantification was set at 10 ng/ml due to a deviation > 10 % of the calculated value to the nominal values by using a linear regression. The relationship between measurement signal and setpoint was no more linear. A change of the surface tension, for example, caused by DNA, can change the curvature of a liquid surface. This changing of the path length in the well of a microtiter plate influenced the measurement signal according to the Beer-Lambert law [162].

In order to confirm the functionality of the detection reaction, the signal of the initial concentrations was compared for each buffer condition. It was shown that the detection signal decreases for lower pH. Furthermore, the effect of salt was more crucial for the detection of calf thymus DNA.

Appendix

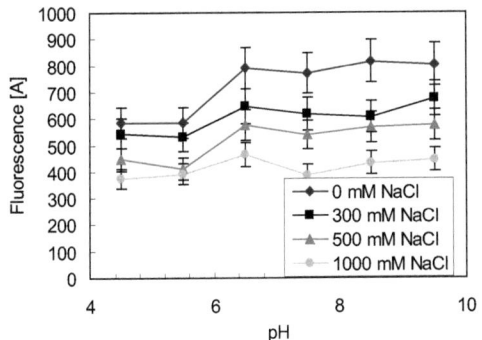

Figure 49: Influence of salt and pH on the PicoGreen® assay for calf thymus DNA. The buffers used are described in Chapter 8.1.2. The DNA concentration was 100 ng/ml.

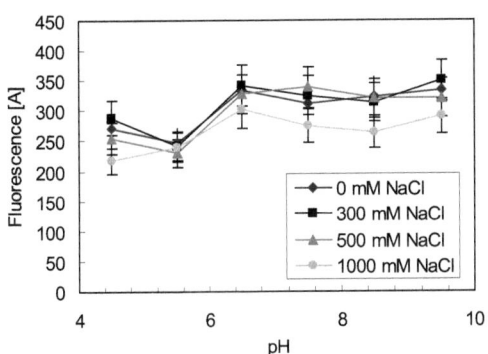

Figure 50: Influence of salt and pH on the PicoGreen® assay for salmon sperm DNA. The buffers used corresponded to those specified in Chapter 8.1.2. The DNA concentration was 100 ng/ml.

8.2.6 Endotoxin assay

The LRV was determined using the kinetic chromogenic limulus amebocyte lysate test according to the manufacturer's instructions. It is based on coloration caused by the contact of the sample containing endotoxin with a mixture of lysate and chromogenic substrate [163]. The rate of formation of the dye is proportional to the concentration of endotoxin.

The equipment in contact with endotoxins was treated with 1 M sodium hydroxide for 30 min or heated at 200 °C for 4 hours. 50 µl of substrate was mixed with the sample in a 96-well half-area plate. Because glucan can interfere with the detection assay, a ß-glucan blocker was added. The kinetic reaction was measured continuously at 405 nm at a temperature of 37 °C using the Tecan plate reader. The required time (onset time) was determined to reach an absorption value of 0.8. By logarithmic calculus a linear relationship between time and endotoxin concentration was found. The log reduction value was calculated similar to the phage assay. The detection limit of the assay was set according to the lowest value of the standard curve. A threefold determination was performed for each sample.

As for the other assays the effect of various buffer conditions was evaluated. In general, for each buffer condition a standard curve should be set up. Thus, during the screening the number of data points increased to very high scores. To circumvent this disadvantage standard curves at different conditions were show in a very similar slope of the linear functions. The presence of salt or a change of the pH changes the rate of the reaction uniformly for different concentrations by only shifting the curves on the x-axis, but not changing the slope

Appendix

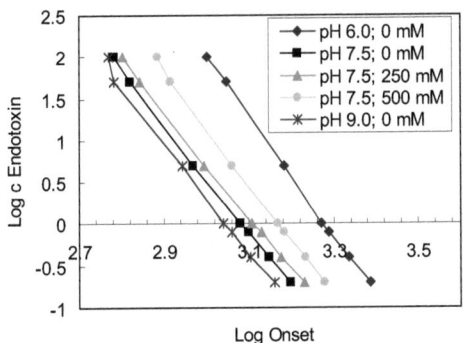

Figure 51: Influence of salt and pH on the endotoxin assay. For different conditions linear standard curves were determined and compared. The endotoxin concentration was between 0.2 and 100 EU/ml. The coefficient of variation of the slopes at different buffer conditions was 4.9 %.

Therefore only a calibration curve at pH 7.5 was established. The samples and dilutions of the initial solutions were analyzed on the same microtiter plate. For the screening studies the initial solution of 500 EU/ml was diluted to 1 EU/ml (Onset$_{Is}$) using the same buffer. By the comparison of the onset time of the calibration curve at 1 EU/ml (Onset$_{STD}$) to the diluted initial concentration the deviation caused by the specific buffer condition could be calculated. Due to the relative change of the reaction time, the measured values (Onset$_{Sample}$) of the samples were corrected to the calibration curve (Onset$_{SampleSTD}$).

$$Onset_{SampleSTD} = Onset_{Sample} + \left(Onset_{Sample} \times \left(\frac{Onset_{STD} - Onset_{Is}}{Onset_{Std}} \right) \right) \quad (8.3)$$

Thus, the concentrations in the flow-through fraction and the LRV were calculated using the calibration curve (20 mM Tris pH 7.5) of each run. The coefficient of variation for the calculated endotoxin concentration was approximately 12 %.

Appendix

8.2.7 Measurements using a FPLC system or a peristaltic pump

For scale-up studies the MA15 device containing of 15 cm² membrane area was used. Protein-binding studies were performed using the Fast Protein Liquid Chromatography (FPLC) system ÄKTA prime. After flushing all tubings of the FPLC, the absorbance of the initial protein was determined by loading 20 ml of the protein solution to the system and through the flow through cell until a stable reading was reached. Then the tubing was flushed again with buffer to reach a 0-reading again. Subsequently, the vented device was connected and flushed with 30 ml buffer. The void volume of the chromatographic system including the membrane adsorber device was detected using 2 % acetone and determined to 2.17 ml ±0.23. The breakthrough curve was recorded at a flow rate of 15 ml/min. The acquired data was exported to an Excel file. The area below the breakthrough curve was integrated to the similar load as in the screening experiments. The binding capacity at a certain loading was calculated by

$$BC = \frac{\left((V_{Load} - V_V) \times UV_{max} - \int_0^{V_{LD}} f(V) dV\right) \times c_i}{UV_{max} \times A} \qquad (8.4)$$

Where V_{Load} is the volume of a specific load, V_V the void volume, UV_{max} the detection signal of the initial concentration, c_i the initial concentration and A the membrane area in the device. The calculation relies on the linearity of the UV signal in the concentration range used which is depicted in Figure 52.

Appendix

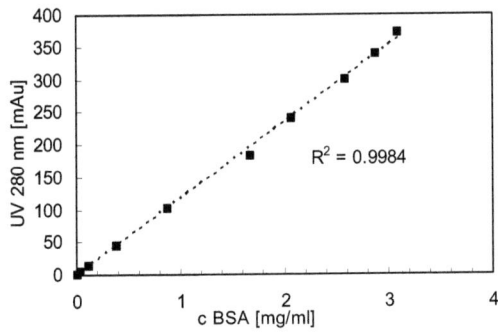

Figure 52: Evaluation of the linearity of the UV 280 nm signal by the ÄKTA prime. Concentrations up to 3 mg/ml were used for the scale-up studies. The linearity was confirmed loading 20 ml of 10 different concentrations ranging from 0.1 to 3 mg/ml to the ÄKTA prime.

For phage studies at a larger scale a multi-channel peristaltic pump was used. The devices were vented according to the manufacturer's recommendations and flushed using 15 ml buffer each. During flushing the flow rate was adjusted to 10 ml/min (corresponding to 24 MV/min). Two devices were filtrated simultaneous. After flushing, 250 ml of the phage solution was loaded to each device. Three fractions of 20 ml were collected after 30, 150 and 230 ml of flow-through. Each fraction was analyzed with the plaque assay. It was found that the determined LRV of all three fractions were within the range of variation for the assay. The constant LRV confirmed that the loaded amount of phages was lower than the expected binding capacity during all phage removal experiments in this thesis. Otherwise a decreasing LRV should have been observed. After each test the tubings of the peristaltic pump were sanitized using 60 ml 1 M/l sodium hydroxide and 200 ml distilled water.

Appendix

8.3 Supplementary results

8.3.1 Determination of volume correction factor and pipetting error

Table 12: Volume correction factor and estimation of pipetting error. The calculation of the volume correction factor (Volcor) and pipetting variation was determined for n = 5 measurements for each tip. The coefficients of variation are given for the 8 pipetting tips. Furthermore

Volume[1]	Characteristics	Tip 1	Tip 2	Tip 3	Tip 4	Tip 5	Tip 6	Tip 7	Tip 8	Tip 1-8
50	Volcor	1.47	2.09	2.22	1.52	1.86	1.81	1.58	1.74	1.79[3]
	Variation[2]	5.29	3.46	3.06	5.29	3.46	2.00	4.16	3.06	5.12[4]
100	Volcor	1.25	1.54	1.61	1.27	1.45	1.37	1.36	1.40	1.41[3]
	Variation[2]	3.46	1.15	3.00	3.21	2.52	1.53	3.00	2.65	2.84[4]
300	Volcor	1.10	1.19	1.23	1.11	1.16	1.16	1.13	1.17	1.16[3]
	Variation[2]	1.07	0.67	0.69	0.51	0.58	0.51	0.58	0.33	0.97[4]
500	Volcor	1.06	1.11	1.14	1.06	1.10	1.09	1.08	1.11	1.09[3]
	Variation[2]	0.12	0.31	0.95	0.90	0.35	0.20	0.20	0.12	0.68[4]
800	Volcor	1.04	1.07	1.08	1.04	1.07	1.06	1.06	1.06	1.06[3]
	Variation[2]	0.38	0.51	0.19	0.19	0.07	0.38	0.14	0.19	0.45[4]
200	Volcor	1.02	1.04	1.05	1.01	1.03	1.04	1.02	1.04	1.03[3]
	Variation[2]	0.68	0.40	0.33	0.30	0.40	0.31	1.89	0.15	0.43[4]
5000	Volcor	1.00	1.00	1.00	1.00	1.00	1.00	1.00	1.00	1.00[3]
	Variation[2]	0.09	0.08	0.16	0.06	0.32	0.09	0.05	0.06	0.52[4]

[1] Pipetted volume [µl] [2] Coefficient of variation [%] [3] Average of all tips [%] [4] Between all tips [%]

Appendix

8.3.2 Separation of DNA and GFP

The following results represent the separation of GFP and DNA at various pH values. The test procedure was explained in Chapter 5.3.2.

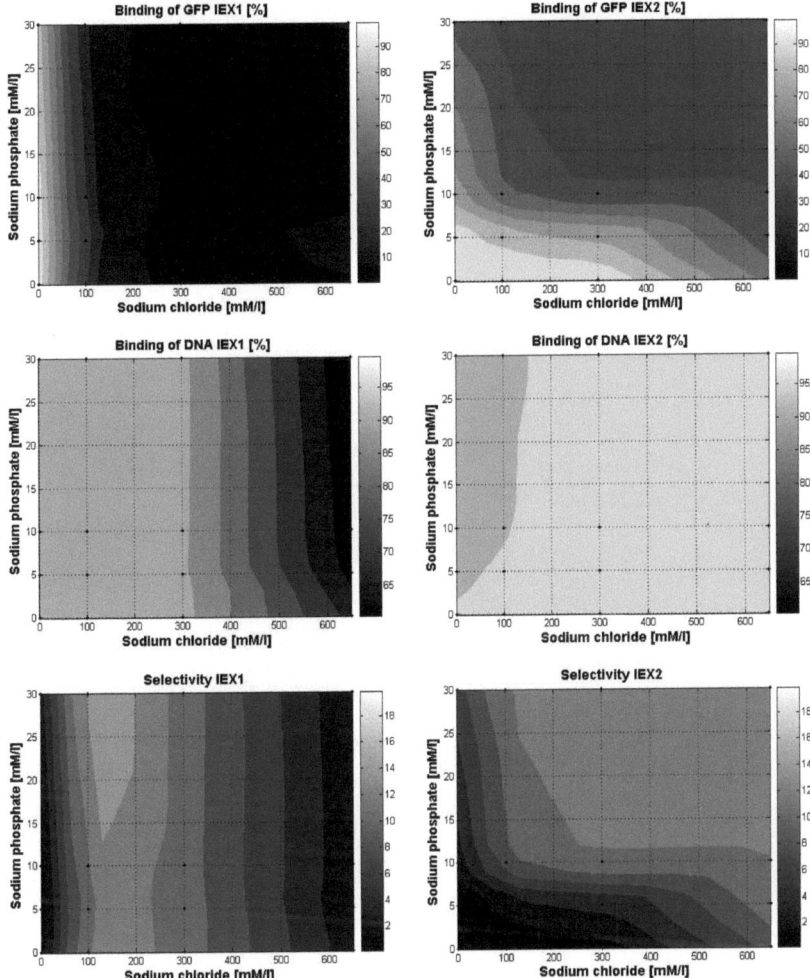

Figure 53: Separation of DNA and GFP at pH 6. Buffer was 20 mM BIS-Tris.

Appendix

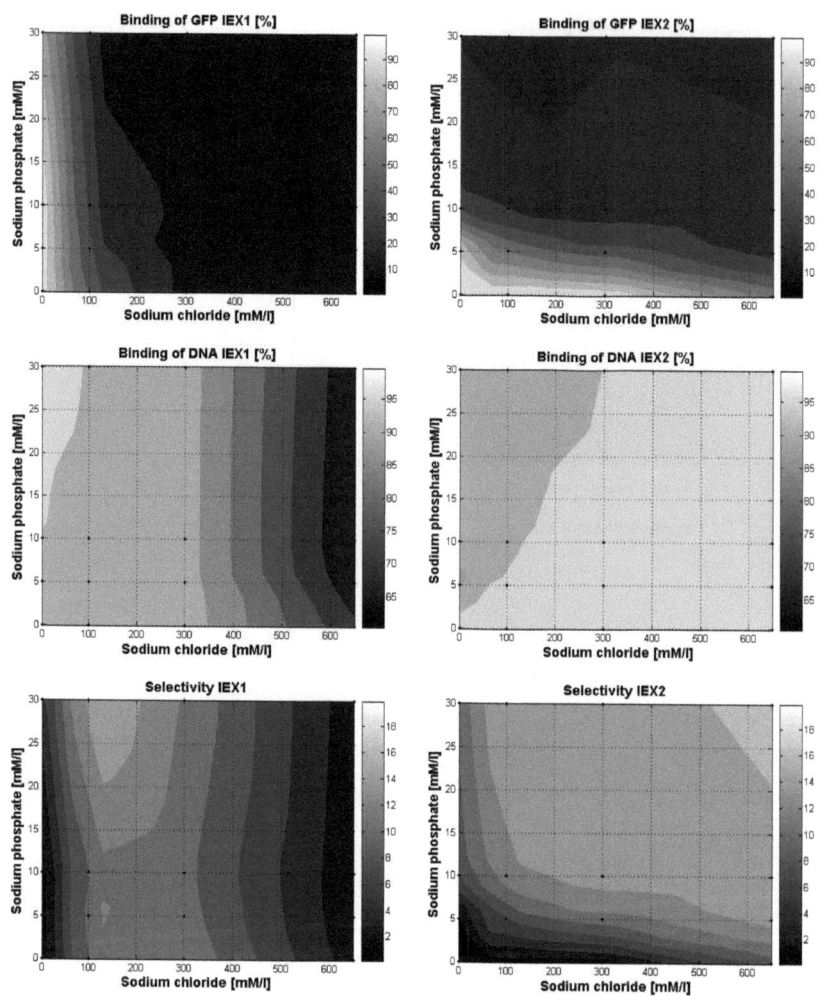

Figure 54: Separation of DNA and GFP at pH 7. Buffer was 20 mM BIS-Tris.

Appendix

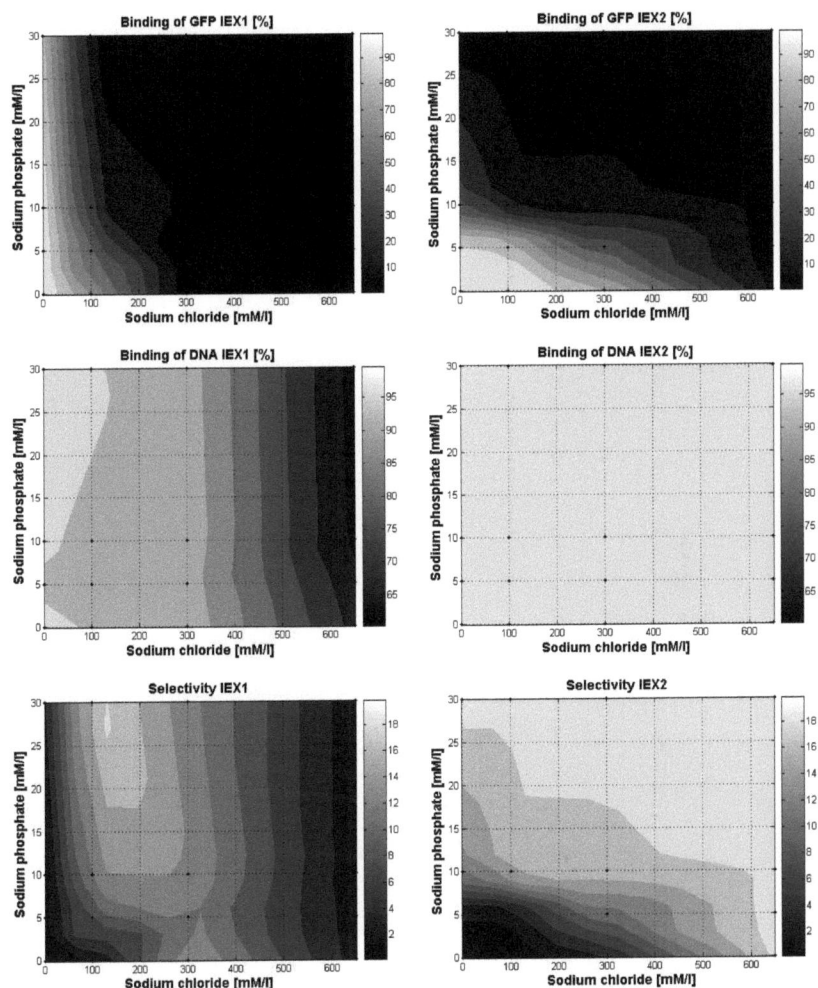

Figure 55: Separation of DNA and GFP at pH 9. Buffer was 20 mM Tris.

i want morebooks!

Buy your books fast and straightforward online - at one of world's fastest growing online book stores! Environmentally sound due to Print-on-Demand technologies.

Buy your books online at
www.get-morebooks.com

Kaufen Sie Ihre Bücher schnell und unkompliziert online – auf einer der am schnellsten wachsenden Buchhandelsplattformen weltweit! Dank Print-On-Demand umwelt- und ressourcenschonend produziert.

Bücher schneller online kaufen
www.morebooks.de

 VDM Verlagsservicegesellschaft mbH
Heinrich-Böcking-Str. 6-8   Telefon: +49 681 3720 174   info@vdm-vsg.de
D - 66121 Saarbrücken   Telefax: +49 681 3720 1749   www.vdm-vsg.de

Printed by Books on Demand GmbH, Norderstedt / Germany